PARTICIPATORY CULTURE, COMMUNITY, AND PLAY

Steve Jones
General Editor

Vol. 75

The Digital Formations series is part of the Peter Lang Media and Communication list.
Every volume is peer reviewed and meets
the highest quality standards for content and production.

PETER LANG
New York • Bern • Frankfurt • Berlin
Brussels • Vienna • Oxford • Warsaw

ADRIENNE L. MASSANARI

PARTICIPATORY CULTURE, COMMUNITY, AND PLAY

learning from reddit

PETER LANG
New York • Bern • Frankfurt • Berlin
Brussels • Vienna • Oxford • Warsaw

Library of Congress Cataloging-in-Publication Data
Massanari, Adrienne L.
Participatory culture, community, and play:
learning from reddit / Adrienne L. Massanari.
pages cm. — (Digital formations; vol. 75)
Includes bibliographical references and index.
1. Online social networks. 2. Online chat groups. 3. Computer bulletin boards.
4. Information technology—Social aspects. I. Title.
HM742.M36   302.30285—dc23   2014044348
ISBN 978-1-4331-2678-9 (hardcover)
ISBN 978-1-4331-2677-2 (paperback)
ISBN 978-1-4539-1501-1 (e-book)
ISSN 1526-3169

Bibliographic information published by **Die Deutsche Nationalbibliothek**.
Die Deutsche Nationalbibliothek lists this publication in the "Deutsche
Nationalbibliografie"; detailed bibliographic data are available
on the Internet at http://dnb.d-nb.de/.

The paper in this book meets the guidelines for permanence and durability
of the Committee on Production Guidelines for Book Longevity
of the Council of Library Resources.

© 2015 Peter Lang Publishing, Inc., New York
29 Broadway, 18th floor, New York, NY 10006
www.peterlang.com

All rights reserved.
Reprint or reproduction, even partially, in all forms such as microfilm,
xerography, microfiche, microcard, and offset strictly prohibited.

Printed in the United States of America

CONTENTS

Acknowledgments	vii
Chapter 1. Introduction	1
Why reddit?	2
Contemporary media cultures and gatekeeping in Web 2.0+	7
A few notes on method	11
Overview of this book	15
Chapter 2. Defining reddit	19
Reddit as carnival	20
Reddit as performance/ritual	21
Reddit as play	22
Reddit as community	23
Reddit as platform	24
What reddit is not	25
Chapter 3. Cynicism, altruism, and hating everything, including ourselves	31
Altruism and its limits	32
Cynicism on reddit: the case of /r/cringe and /r/cringepics	40

Activism and mobilization	45
Pseudoanonymity, authenticity, and the reddit community	50
OP is a faggot	56
How redditors view themselves	58
Rewarding the authentic; punishing the disingenuous	63

Chapter 4. Reliving the past [REPOST] ... 67
Community membership—the transition from
newbie to "redditor" ... 68
Why everyone hates the defaults ... 70
Reddiquette and its discontents ... 73
Reddit's obsession with and repudiation of its own history ... 77
Meta-subreddits ... 84
From patterned interaction to the reddit
circlejerk/hivemind ... 92

Chapter 5. Playing seriously ... 95
Patterns of play ... 95
Karma(ic) rewards ... 114
Reddit rules ... 120
Inverting play ... 122
What reddit can teach us about play ... 125

Chapter 6. Open platforms, closed discourse ... 127
Geek culture and hegemonic masculinities ... 128
Reddit's male gaze ... 130
/r/ShitRedditSays (SRS) and its contrarians ... 143
Where open-source idealism meets reality ... 154

Chapter 7. Conclusion ... 159
The future of reddit ... 161
Revisiting the politics of platforms ... 165
The end of participatory culture? ... 167
Future directions ... 169

Glossary ... 171
References ... 175
Index ... 201

ACKNOWLEDGMENTS

This book has benefited greatly from discussion with my colleagues at the University of Illinois at Chicago, as well as those at other institutions with whom I have had the pleasure of discussing early versions of this work at conferences. Thank you to all of the folks at Peter Lang who have been a pleasure to collaborate with on this process, and especially my editor (and colleague) Steve Jones, who I feel incredibly lucky to work with on a daily basis.

I must express my deepest gratitude to the reddit community, without whom this book would not exist. Thanks especially to all of you willing to answer my interview questions honestly and give me a sense of your experiences with reddit, both the good and the bad.

Also, special thanks to my family and friends, especially my parents, Alice and Jared, who were incredibly supportive throughout the writing process and read early drafts of this book. And, Grandpa Massanari, this one's for you.

· 1 ·

INTRODUCTION

If the internet is made of cats,[1] reddit.com (reddit) is its temple. Like most redditors, I cannot recall precisely how I first heard about the site—just as I cannot imagine my online life without it. I am sure I started visiting it semi-regularly in 2008/2009, mostly for the cat pictures. At the time, I was fascinated by LOLcat humor (Shifman, 2014) and was looking for other spaces online where people shared my fascination. I was pleased to find a group of witty, interesting, like-minded individuals (and some jerks, of course). So basically, I came to reddit because of cats. I stayed for the community.

This is a book about a particular online space—reddit—at a particular time (2011–2014). But more than that, this is a book about how members of online, participatory platforms experience these spaces, how their experiences are shaped by the ways in which they are designed, and how these spaces embody certain contradictions that remain underexplored by scholars and journalists. This book is intended for anyone interested in issues around participatory culture, gender, play, and identity and is meant for both new media scholars and those with just an interest in how we live our lives online now.

This book has two goals. First, it is meant as a critique of much of the work that has already been done concerning participatory culture.[2] I believe that both journalists and academics have often exaggerated the democratic potential and minimized the actual contradictions that it embodies. Popular press accounts of participatory culture tend to be overly laudatory, generic, and reductive. And research into participatory culture platforms often references general and high-level patterns but rarely offers specific details about the way these spaces are experienced, contested, and co-created by both participants and site designers. Additionally, investigations into online participatory culture often do not unpack the ways in which these spaces are contested and negotiated by members—and how "affiliation" (Jenkins, Purushotma, Weigel, Clinton, & Robison, 2009) is neither clear-cut nor separate from the everyday lived experiences of its participants. We tend to gloss over tensions that actually characterize interactions in these communities, including the ways in which members negotiate their own participation, and how the community collectively defines its unique identity in opposition to/connection with other spaces (both online and offline). And we tend to neglect how platform politics (Gillespie, 2010) shapes particular spaces in particular ways—and underplays the importance of how individuals and technologies are co-constitutive (Bijker, Hughes, & Pinch, 2012; Latour, 1992, 2005). To that end, this book focuses on one particular online space, reddit, to engage with some of the larger issues related to mediated, participatory culture in the early days of the 21st century.

Second, by documenting how reddit culture exists now, this book also argues for the importance of preserving the nuances of interactions in these spaces *as they occur*, as global culture is increasingly lived in and enacted through them. However, the lack of meaningful archives makes reconstruction of interactions after the fact nearly impossible, making it critical to document what happens online as it actually happens. It is easy to dismiss "toaster studies" (M. Gray, 2012) approaches to new media—that is, where scholars and journalists examine every new platform or technology superficially—but it is dangerous not to examine these platforms in depth for the richness of the meanings their users make of them. Additionally, I would argue that we should not presume (1) that investigations of the "ephemera" of new media are not critical; or (2) that as scholars (and technology users), we possess the ability to determine what will be "lasting." Doing so risks losing important aspects of our collective cultural heritage (and yes, even LOLcats are a part of that).

Why reddit?

Reddit (the self-proclaimed "front page of the internet") enables the sharing of original and reposted content from around the web. It was founded in June 2005 by Alexis Ohanian; his college friend Steve Huffman sold it to Condé

Nast publications in 2006, and in 2011 it became an independent subsidiary of Condé Nast's parent company, Advance Publications (Ohanian, 2013). In 2008, reddit's code base went open source, becoming a platform that others could use to create their own, entirely separate communities.

Reddit serves as an aggregation platform, which means that most content on the site is linked to rather than directly hosted by reddit. As with other social link/news-sharing sites, registered individuals can submit and upvote or downvote content and comments. Voting is intended to show others what material deserves more (or less) attention from the community. Individuals who submit content and comments are rewarded with karma points, which is a net total of the number of upvotes minus downvotes an item has received. While participants often post content from traditional news organizations, much of reddit's activity revolves around the sharing of original/remixed content, such as memes (/r/AdviceAnimals, /r/ffffffffuuuuuuuuuuuu, /r/lolcats), pictures of animals (/r/aww, /r/babyelephantgifs), solicitations for advice or information (/r/AskReddit, /r/askscience, /r/explainlikeimfive, /r/IAmA, etc.), or niche interests (/r/PenmanshipPorn, /r/hiphopheads, /r/bicycling, /r/MakeupAddiction) in both public and private subreddits.[3] Figure 1.1 shows what the default front page (/r/all) might look like for a non-member.

Figure 1.1. Front page of reddit.com/r/all for a non-member (captured November 5, 2014).

Reddit members ("redditors") can customize their home page to see only subreddits of interest. Thus, each individual's experience of the site is a bit different, depending on his or her subscriptions. However, a number of subreddits serve as "defaults"—meaning that new accounts are automatically subscribed to them.[4] The intention of the defaults is to showcase to new users a variety of the content that the site offers—the reality is somewhat different (see Chapter 4).

Figure 1.2. Reddit front page for logged-in user (captured November 5, 2014).

While the links are interesting, most of what makes reddit an engaging place are the discussions around submitted content. Clicking on the smaller comment link instead of the main link takes you to the discussion page for that particular image, video, article, or text posting. (See Figure 1.3.)

Here is where the best (or worst) of reddit is on display. Like links and text postings, comments can be up- or downvoted by redditors, with those that have the most upvotes typically appearing closer to the top of the page. The reddit platform also allows for minor formatting in comment threads—including links, tables, bulleted lists, and blockquotes.[5]

Reddit also features rudimentary profile pages where you can view links and comments submitted by an individual account as well as see a running

Figure 1.3. Cropped version of a comment page (captured November 5, 2014).

total of that individual's comment and link karma (useful tools for determining whether he/she is a troll). Profile pages also feature a "trophy case" with badges indicating the length of time the account has existed, any participation in Secret Santa exchanges, Reddit Gold status, and other awards such as "best comment" that the account may have received.[6] Redditors can also send private messages (PMs) to others, as well as "friend" them, which basically just highlights their username on link listings and in comment threads.[7] (See Figure 1.4.)

While scholars have considered user participation and information sharing on other, similar Web 2.0 (O'Reilly, 2005)/participatory culture platforms such as Digg, delicious, and Metafilter (Lerman & Ghos, 2010; Marlow, Naaman, Boyd, & Davis, 2006; Silva, Goel, & Mousavidin, 2009), little research focuses specifically on the reddit community, despite its growing popularity. As of July 2014, reddit was the 19th most visited site in the US and 54th globally ("reddit.com," 2014). Visitors to reddit spend about 13 minutes on the site ("reddit.com," 2014) and during June 2014, it had over 114,540,000 unique visitors from over 190 countries ("About reddit," 2014). Additionally,

Figure 1.4. Cropped version of a profile page (captured November 5, 2014).

mainstream media outlets have begun to acknowledge its importance as an engaged—and, at times, problematic—community (Banks, 2013; Chen, 2012; Kang, 2013; Zuckerman, 2013). Reddit's political activism around attempts to limit net neutrality and curtail internet freedoms in the US, as well as the site's involvement with the Boston Marathon Bombing investigation in 2013, highlight its potential for enabling collective action, whether for good or for ill. Frequent appearances by politicians, actors, writers, scientists, and public intellectuals on the site's /r/IAmA subreddit have also raised the profile of the site. At the same time, its reputation as an "anything goes" space has led to significant ethical and legal questions regarding content—from racist subreddits such as /r/GreatApes, to sexualized content featuring minors and upskirt photos of women on the now-deceased /r/creepshots.

While interaction on reddit is necessarily social, it is not, strictly speaking, a social networking site where connections are "publically articulated" (Ellison & boyd, 2013). Facebook, Google+, and the like tend to rely on one's online identity connecting to one's "offline" identity—or at least that is increasingly becoming their conceit. And, unlike YouTube, Vimeo, or Soundcloud, reddit is not organized around a particular media artifact type (such as videos or sound clips). Nor does interaction revolve exclusively around niche discussions of content (such as MetaFilter or Slashdot or discussion boards generally). Nor is interaction managed and controlled by codified individuals or groups (such as on blogs). And unlike its anonymous close cousin 4chan, reddit is a pseudo-anonymous, persistent space, where individuals gain reputation and recognition over time. Reddit is therefore a unique, boundary-spanning platform that elicits new questions about the nature of participatory culture and community in the age of social networking.

Contemporary media cultures and gatekeeping in Web 2.0+

It is impossible to discuss a site like reddit without first noting the ways in which media cultures are changing in the 21st century. Henry Jenkins's (2006) idea of "participatory culture" suggests that our individual and collective engagement with mass media fundamentally shifted during the early 1990s, around the same time that the internet became more commonly available in people's workplaces and homes. However, it is important to acknowledge the sort of active engagement we see today with popular media is not new—fanzines, underground publications, and the like have been around since the 1940s (and amateur press and radio involvement long before that)—but the scale and pervasiveness of our ability to "talk back" to the media, and, most important, be heard by media producers, experienced a rapid increase concomitantly with the internet's penetration into our everyday lives (Jenkins, Ford, & Green, 2013). As audiences became fractured, niche, and increasingly invested in the media products they "consumed," the tools used to create content became cheaper and more accessible to the average fan. Thus, everyone potentially can become a "prosumer" of content (Jenkins, 2006). This shift—from a massified audience that was perceived as being composed of relatively passive receptors in the media production chain to a more active and engaged audience whose members produce media content of their own—has been well documented (Andrejevic, 2008; Baym, 2000; Delwiche & Henderson, 2013; Murray, 2004; Shefrin, 2004).

We increasingly view audiences as active participants who "work for the text" (Milner, 2009) and are willing to labor freely for non-monetary gains such as social connections with other fans and deeper engagement with the texts themselves. And participatory culture platforms fundamentally depend on user-generated content. One would think this would result in a relatively symmetrical relationship between platform owners and their audiences, as the former is dependent on the latter's continued engagement and willingness to produce new content. Critics of the neoliberalism underpinnings of Web 2.0 (and its later iterations) have argued, however, that "free" labor can be exploited labor (Kosnik, 2013; Ross, 2013; Scholz, 2008, 2013; Terranova, 2003). For example, platform owners extract value from users in the form of data and metrics, creating in essence algorithmic "taste profiles." These are then sold to companies for marketing purposes, which are later used to create targeted advertisements for users (van Dijck, 2013). Both Jaron Lanier (2010) and Evgeny Morozov (2013) note the implicit irony of organizations harvesting and essentially selling back the collective labor of the individuals without which the platform would cease to exist (or at least cease to have much relevance or market penetration).

In addition to changing the logics between producers and consumers, the internet in general and sites like reddit specifically present a challenge to media gatekeeping. Traditionally, news organizations have relied on specialized professionals (journalists and editors) to ensure content is ethically collected and reported. In the mass media environment of the early to mid-20th century, this meant that a small number of individuals (gatekeepers) were responsible for determining what constituted "the news" (Bennett, 2004). It also meant that news organizations frequently depended on elites for access, creating a cycle whereby newsworthiness was partially determined by what elites wanted the public to know. And what news organizations covered became part of the national dialogue—that is, news coverage partially determined what the public would talk about (McCombs & Shaw, 1972).

The development of Web 2.0 technologies altered this relationship between news organizations and the public. It encouraged a rise in citizen journalism, whereby non-professionals were able to participate in pursuing news stories of interest, which resulted in a proliferation of "non-official" sources (Gillmor, 2006). This meant that gatekeepers no longer solely controlled the dialogue between the populace and the elites. In the mid-2000s, the growing popularity of social networking platforms meant that citizens could reach out directly to elites through these channels, reducing the need for—or at least the appearance of a need for—intermediaries between the two groups. So the populace had both more access to elites and additional channels of information beyond the "mass media" of the previous century. Theoretically, this would mean that the public would be better informed by a greater diversity of news sources than ever before. Civic and political participation would naturally rise as a result of these changes (Norris, 2001).

The reddit community's desire to break down the barriers between expert and novice is evidenced in the large number of subreddits dedicated to non-experts asking questions of scientists (/r/askscience) or historians (/r/AskHistorians) and blurs the lines between public and private figure. The community's effective meme-spreading abilities have also catapulted private individuals to internet and popular media glory. See, for example, Zeddie Little, who gained fame on reddit as "Ridiculously Photogenic Guy" (Brad, 2012a). Likewise, popular posters on the site, such as biologist /u/Unidan and his trademark "Biologist here" opening line, become micro-celebrities themselves (Marwick, 2013; Senft, 2013). And the site's hugely popular default subreddit /r/IAmA hosts conversations with celebrities, politicians, actors, authors, and scientists and allows the average redditor to ask a question of or have a short conversation with prominent cultural figures directly. The implication is that since there are no intermediaries (or at least no overtly visible ones), the conversation occurring is more democratic, more authentic, and more deliberative.

In practice, however, the factor that attracts many redditors to the site—the ability to create a feed based on individual interests, no matter how niche—also contributes to the reality that it is easy to create an "echo chamber" or "filter bubble" (Pariser, 2011). The sense that content on the site is completely tailored to your interests and regularly refreshed makes it both an intoxicating space and incredibly addicting. But it also means that it is easy to create an experience wherein one's views are merely confirmed rather than challenged, likely the opposite of the democratic, deliberative nature of the "potential" of the internet. Most redditors are aware of these twin tendencies. As Dakota[8] told me:

> There's a near-constant stream of new, interesting content … And once you curate your subreddits to your liking, almost all of the content is directly appealing to you … In addition to selecting your subreddits by content, you're probably also sorting them by conversation—it's more than possible to only subscribe to subs [subreddits] such that you converse almost exclusively with scientists, or Joss Whedon fans, or people of your own language (if it's not English). It's really easy to create your own "bubble" of people who think and speak like you do—they may challenge your ideas at times, but they're otherwise your "peers." I think that's an especially appealing/"addictive" aspect, because, as an adult with[out] the social constructs of school, it can be challenging to create such communities IRL.

As this individual articulates, redditors are aware of the reality of having a limited experience of the site's full content while still liking it for precisely these reasons. The "bubble" that my informant mentions is one that many redditors happily exist inside—even while threads decry the "echo chamber" or "hivemind" nature of many of the site's conversations. It is not unsurprising, then, that experimental research on sites with similar downvoting/upvoting mechanics as reddit demonstrates that social influence plays a key role in how postings are received (Muchnik, Aral, & Taylor, 2013, p. 650). And yet redditors remain a news-consuming bunch: 62% of them get news from the site (as compared with 52% for Twitter, 47% for Facebook, and 20% for YouTube) (Holcomb, Gottfried, & Mitchell, 2013).

However, the notion that reddit is entirely without gatekeepers is untrue—or at least the reality is more complicated. Moderators have a great deal of power over the kind of content that appears in their subreddits. They can set explicit rules regarding sources, tone, or content of submissions and comments. Yes, anyone can create his/her own subreddit, but actually creating one that gains subscribers and visibility is tricky. A second set of "gatekeepers" is composed of those individuals who browse the new queue of any particular subreddit and vote on the most recent submissions. The reddit ranking algorithm weighs the earliest votes on a particular item or comment more heavily, meaning that the first 10 votes are most influential in terms of whether other redditors will ever see the submitted material (Salihefendic, 2010). This informal, anonymous group is referred to fondly as the "Knights of the New" by redditors. This term of endearment is applied because the task is seen as onerous and mind numbing (a lot of spam and meaningless content clutter up the new queue at any given moment) and because redditors recognize the importance of preserving the quality of the site submissions. So these individuals also act as kind of ad hoc gatekeepers as well. In fact, I suspect that many of the Knights of the New are also moderators—as they would have both the inclination for and interest in keeping content quality up while also spending a significant amount of time on the site because of their moderation duties. The notion that a select few individuals might control much of what becomes popular is discussed regularly on the site—most notably in a /r/TheoryofReddit posting in which some redditors decried the "power user" model they believed led to the downfall of Digg.com. As the original poster (OP) wrote, "The trend in the past year seems to be toward a centralization of power (and we all know power has a rather unfortunate side-effect of corruption, especially on the Net), reduction of mod accountability, and painting any criticism as 'rabble rousing' or 'witch hunting'" (smooshie, 2012). Thus, reddit represents a complex space in which the influence of elite users and gatekeepers may be reduced but is not eliminated entirely.

A few notes on method

This research is shaped by the ethnographic approaches that other scholars have used to understand communities enabled by the web and other new media (Beaulieu, 2004; Boellstorff, Nardi, Pearce, & Taylor, 2012; Hine, 2008; Horst & Miller, 2012; Markham & Baym, 2009). Ethnography ("writing culture") attempts to unpack the culture of a space and explore it from the perspective of an insider (Clifford & Marcus, 1986). Using a methodological orientation rather than simply a single method, ethnographers employ a number of approaches to understanding cultures, including participatory observation, interviews, text analysis, and historical research (Schensul & LeCompte, 2013), all while taking copious, detailed field notes (Emerson, Fretz, & Shaw, 2011; Van Maanen, 2011). Digital, online, or internet ethnography applies these methods online, as an attempt to understand how individuals, communities, and cultures shape and are shaped by these spaces.

Qualitative researchers and in particular those inclined toward ethnographic methods, emphasize the importance of reflexivity within the research process. Annette Markham (2009), for example, argues that it is critical for researchers to acknowledge their local subjectivities and encourages them to reflect on the ways in which their interpretations are colored by cultural assumptions. To that end, I offer a few notes about my involvement with reddit. First of all, this community feels familiar. I worked for a number of years as an HTML coder (when such a thing existed), user interface designer, usability expert, information architect, and all-around "webmonkey." My workplaces were overwhelmingly young, nerdy, white, male dominated, and liberal. I have also long been interested in elements of geek culture—as an avid video gamer, lover of independent comics, and design nerd. Most important, I love cats and reddit does too. Therefore, reddit seems like home to me, even if I find myself frustrated by the lack of diversity and problematic discourse on a site that rhetorically espouses the importance of free speech, informed opinions, and egalitarianism.

As of August 2014, my primary account has been registered for only three years, but I lurked for several years before that. This actually seems to be a common pattern for reddit (and, possibly, most pseudoanonymous communities where the stakes of contributing versus lurking are relatively high)—many posters make their first posts about the de-lurking process and often attempt to inoculate themselves from criticism by posting the words "first post; please be kind" or the equivalent. The process by which I de-lurked was uneventful—I posted several pictures from my travels in Hong Kong in 2011. More specifically, I submitted a blurry picture I took in a Hong Kong mall of a Mona Lisa eating an ice cream cone which was entirely made of miniature toasts. My first post did not hit the front page of the subreddit to which it was submitted, nor did it go entirely unnoticed—I earned a paltry 25 karma for it. At the time, I was sure that the content would be of interest to others on the site if I could only come up with a pithy headline to accompany it—or if it had actually been taken with a real camera rather than my terrible cell phone.

Since my first posting on reddit, I have upvoted and downvoted thousands of posts and comments, commented on others' submissions, conducted interviews with other reddit members and moderators about their experiences with the site, and collected almost anything written about reddit—from official historical documents that trace the founding of reddit to observations and criticism written by journalists and bloggers. Some interviewees/informants were recruited through a subreddit specifically dedicated to such requests (/r/SampleSize), while others, such as the moderators of ShitRedditSays (SRS), were contacted directly because of their unique perspective on the site. Interviews helped me fill in the gaps about how others came to reddit, how they understood the space, what they found enjoyable about their experiences, and what they wished was different about it.

I have also stayed quiet and simply watched interactions on the site unfold. My corpus includes copious screenshots, written field notes, and coded interactions in an attempt to understand the particularities of reddit's culture. Much of this research is grounded in my observations of reddit's generic front page (/r/all) and the more popular subreddits. This is not to suggest that reddit's collective identity is only shaped by the few subreddits whose postings seem to dominate the front page. Instead, it is merely a convenient starting point for grappling with the enormous quantity of original content that reddit members create.

Reddit's platform logics add a layer of complexity to this project. Posters are encouraged to upvote content they find useful, interesting, or that adds meaningfully to the conversation according to the site's reddiquette.[9] Both comments and postings are then moved to the top of the page based on the number of upvotes minus the number of downvotes they receive.[10] A large amount of karma for a particular comment may indicate that individuals reading it found it valuable or agreed with it or thought that it added useful information to the conversation or thought that it was funny—in other words, it is difficult to know what might be motivating an individual redditor's vote. In addition, comments that have received a large number of downvotes are often hidden, depending on the threshold a user sets in the preferences screen, meaning that a significant portion of the reddit audience may never see them.[11] In this project, therefore, comments and posts with a large amount of karma were viewed with a bit of skepticism and not used as the sole mechanism for understanding reddit culture. As with other communities, popular commenters and perspectives tend to emerge, creating a sort of power law dynamic (Shirky, 2003) in which a significant portion of the content and comments tends to be ignored in favor of those made by a few redditors (karmarank, 2014). Still, it is important to note what content is favored by the reddit community and what is not—even if the motivation behind a particular up- or downvote is not known.

Ethnographers have long discussed the "loss of distance" that happens when engaging with a community. My experience with reddit is no different. On a daily basis, I see material that makes me think, makes me laugh, and touches my heart. Many of these moments are ineffable and fleeting and unfortunately dismissed as meaningless for many "outsiders." At the same time, there are moments—too many to count—during which my jaw drops open, I sigh heavily, and slam my laptop shut after reading yet another misogynistic, ablest, racist, homophobic, ageist, transphobic posting or comment. Reddit is regularly an infuriating *and* inspiring place. Doing ethnography is complicated; cultures are messy, slippery, and complicated things that can rarely be summed up in tidy packages. Ambiguity and qualifying statements seem to be the norm in many ethnographies, and this one is no different. This ethnography of reddit is literally one story of potentially many that could be told about this space. It is, however, a narrative grounded in many different forms of data and is one that I feel confident will at least serve as a starting point for creating a more nuanced understanding of how participatory culture platforms are multifaceted places embodying democratic ideals while still marginalizing certain voices.

Critics could question my attempt to characterize reddit as *a culture* rather than simply an assemblage of *many cultures*. It is true that reddit is not a unitary entity. It is a space that hosts a wide variety of subreddits, each of which has its own unique subcultures, norms, and rules. But it is also true that certain patterns characterize interactions throughout the site. Some of these are a function of the reddit voting system, while others reflect larger demographic tendencies and attitudes of the reddit user base. So acknowledging that reddit is composed of many different subcultures, some of which hold diametrically conflicting perspectives, does not negate the importance of understanding how larger cultural mores shape the space. I realize this approach is likely to be met with some resistance by some within the reddit community. Redditors often push back against the argument that reddit has "a culture," instead suggesting that it reflects many different cultures while at the same time decrying reddit's "hivemind." However, this does not reflect my own experience or that of at least some others within the community. For example, one individual on /r/circlebroke (a subreddit dedicated to meta-conversations about reddit with a social justice perspective) argued that even if redditors are not willing to acknowledge publically their own cultural baggage, it is reflected in their voting behavior:[12]

> Usually what will happen is someone will criticize Reddit as a whole for one of it's many screwed up views, and no matter what, **every single fucking time**, some jerk-off will smugly respond with something along the lines of:
>
> *It's almost as if Reddit is made up of millions of people who all have different opinions. (+10,000,000 upvotes)*
>
> Every time I see this comment I want [to] punch a hole through my monitor. Do these people not understand that when these circlejerk comments (eugenics, fat shaming, slut shaming, racism, etc) constantly get upvoted by thousands of people that it's probably the general opinion held by the majority of the community? (Frank_Reynolds_19, 2014)

This is one of those both/and situations: reddit is both one and many cultures, but getting a handle on the latter requires a good understanding of the former. And, to alleviate the concern that I am painting this space with too broad a brush, my goal with this book is to highlight rather than resolve the contradictions that I see expressed on reddit and how these reflect larger tensions within mediated participatory culture more broadly. My hope is that this book *begins* a conversation—about the politics of participatory culture platforms, about reddit, and about the need to engage with documenting and understanding contemporaneous studies of online spaces before they disappear.

Overview of this book

This book is organized into themes that highlight the complex and often contradictory nature of reddit. The diversity of subreddits dedicated to niche interests would suggest that nothing really binds this community together; however, it is the unspoken politics of the reddit platform (Gillespie, 2010) that encourages particular interactional patterns across the space. These interactional patterns highlight larger contradictions within the community and serve to deepen our assumptions about the nature of participatory culture. These contradictions include the tendency for the community to be intensely altruistic and cynical with regard to institutions, communities, individuals, and themselves; the tension between new members and longtime "redditors"; the way the community and the platform itself encourages a playful approach to discourse and how this is also contested; and how reddit members and designers espouse a technologically open but discursively closed approach to the community. Each of these tensions/contradictions is explored in its own chapter.

In Chapter 2, I highlight the multifaceted nature of reddit and explain some of the bodies of literature that inform how I approached thinking about reddit culture. This chapter introduces a number of ways to think about the space—as a kind of carnival, a performance, play, community, and as a platform. I also explore how reddit is related to and different from other participatory culture platforms (Twitter, 4chan, YouTube, etc.).

Chapter 3 explores the ways in which redditors relate to themselves and the community both inside and outside of reddit. I describe a number of ways in which the community expresses both altruistic and cynical tendencies. Altruism is evidenced through the numerous social support subreddits, offers of material goods through gift exchanges, and individual gifts of Reddit Gold and cryptocurrency. In addition, the community often rallies around important causes, especially those of interest to its tech-savvy members. As on other online platforms, however, redditors can exhibit a kind of mob mentality—which is evidenced by its involvement in singular events (such as the Boston Marathon Bombing) and also colors much of the discourse on particular subreddits (such as those dedicated to "cringeworthy" postings). I end with a description of the multiple and contradictory ways that redditors think about themselves and their community.

In Chapter 4, I examine the way novice reddit members become "redditors" and how this changes their relationship with the community as a whole.

I also explore the conflicted relationship the reddit community members have with their own history. I focus on the ways in which subreddits tend to drift over time, as well as redditors' obsession with reposting material and their belief that an influx of new users contributes to the decline in content and discourse on the site. In addition, this chapter illustrates how the default subreddits, although the most visible to site visitors, are viewed with scorn by many. It also details the ways that reddiquette—an informal statement of community rules—is both useful and disregarded regularly by the community. I end with a discussion of meta-subreddits—the spaces where reddit discusses reddit—and why the reddit platform and community may encourage these kinds of conversations.

Chapter 5 borrows work from game and play studies to examine the ways in which redditors engage in various forms of play as a primary mode of expression. From memes to novelty accounts to reaction GIFs, I explain how these forms are uniquely enabled by the reddit platform. I also explore how redditors are motivated by karma, how it is publically derided but still holds importance for many individuals. I explain how accusations of "shitposting" and "karmawhoring" suggest that the community values meaningful contributions over that which might be merely popular or repetitive. This chapter also explores the ways the community defines the "rules" of reddit and how they may be inverted. I briefly explain how this kind of inversion of play is different from trolling as a practice.

In Chapter 6, I detail the problematic ways in which reddit discourse marginalizes many voices. Focusing specifically on the ways in which redditors create a kind of "male gaze," I examine how geek masculinity discursively excludes women and people of color—despite their longtime presence in the community. I explore the image of women on reddit in multiple ways—from the NSFW (not safe for work) subreddits to men's rights activism, through memes shared on /r/AdviceAnimals and elsewhere, which often feature a misogynistic and retrograde vision of women. This chapter profiles a specific reddit community—ShitRedditSays (SRS)—that engages in a kind of counterplay in an effort to raise awareness about the problematic and offensive discourse that populates much of reddit. SRS's role as "redditors" is a contested one, and this brings larger questions about the nature of insiders and outsiders in participatory culture to the foreground.

Chapter 7 offers a summary of findings and connects them to a broader discussion of platform politics and participatory culture. It also explores how reddit might develop in the future and offers a brief examination of how pseudoanonymity shapes the site's discourse. I conclude with a discussion of future directions that research into reddit might take.

Notes

1. https://www.youtube.com/watch?v=zi8VTeDHjcM
2. From the MacArthur Foundation's *Confronting the Challenges of Participatory Culture: Media Education for the 21st Century*: "A participatory culture is a culture with relatively low barriers to artistic expression and civic engagement, strong support for creating and sharing one's creations, and some type of informal mentorship whereby what is known by the most experienced is passed along to novices. A participatory culture is also one in which members believe their contributions matter, and feel some degree of social connection with one another (at the least they care what other people think about what they have created)" (Jenkins et al., 2009, p. 3).
3. For the purposes of readability, I have shortened the URLs for the subreddits I mentioned in this text to "/r/nameofsubreddit." Likewise, usernames have been shortened to "/u/nameofuser." Reddit (and redditors) appears lowercase, except when referred to differently by interviewees and press pieces about the site. Subreddit names preserve the capitalization from the site's subreddit list at http://www.reddit.com/subreddits/. And, as a matter of practicality, my work on reddit focused exclusively on the site's public subreddits.
4. The defaults ballooned in May 2014, going from 25 to 50. This list includes /r/announcements, /r/Art, /r/AskReddit, /r/askscience, /r/aww, /r/blog, /r/books, /r/creepy, /r/dataisbeautiful, /r/DIY, /r/Documentaries, /r/EarthPorn, /r/explainlikeimfive, /r/Fitness, /r/food, /r/funny, /r/Futurology, /r/gadgets, /r/gaming, /r/GetMotivated, /r/gifs, /r/history, /r/IAmA, /r/InternetIsBeautiful, /r/Jokes, /r/LifeProTips, /r/listentothis, /r/mildlyinteresting, /r/movies, /r/Music, /r/news, /r/nosleep, /r/nottheonion, /r/oldschoolcool, /r/personalfinance, /r/philosophy, /r/photoshopbattles, /r/pics, /r/science, /r/Showerthoughts, /r/space, /r/sports, /r/television, /r/tifu, /r/todayilearned, /r/TwoXChromosomes, /r/UpliftingNews, /r/videos, /r/worldnews, /r/writingprompts (Hern, 2014).
5. Reddit uses a modified version of Markdown syntax: http://daringfireball.net/projects/markdown/syntax.
6. A list of trophies is located at http://www.reddit.com/wiki/awards.
7. A complete guide describing the basics of reddit can be found at http://www.reddit.com/wiki/reddit_101.
8. All informants/interviewees were known to me only by their reddit username. I have assigned gender-neutral pseudonyms for each interviewee. These were generated randomly from http://www.babynames1000.com/gender-neutral/.
9. Reddiquette is a community-authored statement of general "rules" for reddit behavior, which I discuss more in depth in Chapter 4.
10. The algorithm for showing the totals of upvotes and downvotes (karma) is slightly more complicated than this. Reddit code "fuzzes" the points of a posting to account for spam and does not show the score of newly posted items for the first few hours to avoid a "bandwagon" effect (cupcake1713, 2014a). In June 2014, reddit administrators also made a significant

change to the way in which votes were displayed on the site. Instead of showing the individual total number of upvotes and downvotes, the platform would now display a percentage (initially, Reddit Enhancement Suite users saw a ?|? in which the vote totals had been shown previously—later this was removed entirely). Administrators argued that since the vote totals were already inaccurate, having been fuzzed to avoid providing feedback to spam bots regarding their effectiveness, it was more of a "tweak" than a major change (Deimorz, 2014b). But it proved highly controversial, with many redditors arguing that a link or comment with a score of –1 based on 299 upvotes and 300 downvotes was substantially different from one that had a –1 based on only 2 upvotes and 3 downvotes. The fieldwork and findings reported in this book do not reflect this change regarding how karma is reported to reddit users.

11. Comments on a particular posting can be sorted a number of ways through a drop-down menu on each posting's page. Each of these sorting functions changes the visibility of comments. The default sort function is "best" (which sorts the comments based on the number of upvotes and downvotes, taking into account sample size) (Munroe, 2009). Several other sorting options exist: "top" (this makes comments with the most upvotes, regardless of downvotes, the most visible/highest on the page), "new" (this moves the newest comments to the top of the page), "hot" (this is a combination of the "top" and "new" function, giving some priority to comments that are most recently made), "controversial" (this moves comments with a large number of upvotes *and* downvotes to the top of the page), and "old" (this moves the oldest comments to the top of the page) (djloreddit, 2014).

12. Quotations from informants and public postings retain their original grammar, capitalization, and usage, and have only been edited for clarity where necessary.

· 2 ·

DEFINING REDDIT

When asked what I was working on during this book project, I would simply say, "Oh, it's a book about this online space, reddit." Inevitably, the person I was talking to would respond in one of two ways: nod knowingly and then launch into her own story about her experience with the site, or, much more likely, ask innocently, "What's reddit?" I would then reply with one of my stock answers, "It's a social news-sharing site," or "It's a link aggregator," or "It's an online community." If the person still seemed interested, I would clarify further: "Well, it's not just a social news-sharing site, it's also a community. Anyone can create what's called a subreddit and share links of interest. Like Swedish death metal? There's a subreddit for that (/r/melodicdeathmetal) and people who want to talk about it. Interested in calligraphy? There's a subreddit for that, too (/r/Calligraphy). Do you drink beer in the shower? Awesome—there's also a subreddit for that (/r/showerbeer). But the really interesting stuff happens in the discussions around those links. It's kind of like a community, message board, carnival, and play space rolled into one. Oh, and yeah—you should know there's some really disturbing stuff too. It's kind of like the best and worst parts of the internet and humanity rolled up into one space." And, we would go from there—with me relying on all kinds of metaphors to try to describe elements of the phenomenological experience of being on reddit, and mostly failing to get at what this space is about. These often changed depending on what stage I was in during my research. But several stuck with me: reddit as carnival, as performance, as play, as community, and as platform. Each of these metaphors provided a useful "lens" for viewing the site and offered a set of theoretical literature that informed my ethnographic work.

Reddit as carnival

While reddit consists of many different moderated subreddits, all of which have their own explicit and implicit rules around the kind of content they accept and the interactions they encourage/allow, the idea of "carnival" is a productive lens for understanding site interactions. In Mikhail Bakhtin's (1984) formulation, carnival represents an upending of the trappings of everyday life. Hierarchies are ameliorated and reversed, and the crowd takes pleasure in the "grotesque body," reveling in the pleasures (and degradation) of the flesh. On reddit, this means that a single thread may contain sexual references, animated GIF responses, puns, grotesque images and stories, racist and sexist speech, all juxtaposed with sincere commentary and meaningful dialogue. All of this contributes to a chaotic space that is at times both compelling and repulsive.

As Stuart Tannock (1995) argues, the carnival is a *sanctioned* activity. Those in power encourage or allow the carnival to take place, perhaps in the unspoken hope that the populace will not question the hierarchies that require such a sanctioning to occur. In the case of reddit, the carnivalesque spirit depends on the site's administrators and moderators to allow these kinds of interactions to occur. As with many other internet platforms, the governance structure of the site is relatively flat, and its administrators are mostly "hands off" with regard to the content and tone of the community's many conversations. However, volunteers from the community moderate individual subreddits, serving a critical function in setting the tone of the interactions that occur therein and buffering administrators from users of the site and the content, however objectionable, that occurs.[1] These layers remain relatively transparent to most users of the site, especially those that do not submit content regularly or volunteer to moderate. Revelers may be caught up in the reddit experience, but the party only exists (so to speak) because it is being sanctioned and allowed by the administrators.

At the same time, there can be no carnival without its reverse. As such, it is an implicit critique of the accepted norms governing everyday life. Redditors reveling in the grotesque, crowning the fool the king, and demonstrating a great disdain and irreverence for hierarchies—all are part of the carnivalesque. At times, this irreverence is laudable, when an institution such as the US government or celebrity culture is the target. But the bacchanalia is not always so clear cut. Oftentimes the reddit carnival is a mask for cruelty and mob justice—a point I return to in the next chapter.

Reddit as performance/ritual

A slightly different but related perspective on the reddit experience and space draws on the work of performance studies to understand the ways in which redditors engage in a ritualized performance of community. The work of Victor Turner (V. Turner, 1987, 2001) and that of Richard Schechner (1985) highlight the sociological importance of rituals within cultures and the performative nature of social interactions. Both scholars emphasize the liminal nature of ritualized performance. In reflecting on Turner's impact on the world of performance studies, Schechner writes, "He taught that there was a continual, dynamic process linking performative behavior—art, sports, ritual, play—with social and ethical structure: the way people think about and organize their lives and specify individual and group values" (V. Turner, 1987, p. 8). Thus, rituals reflect cultural values, and cultures are inscribed/enacted through the ritualized performances that its members create. On reddit, this means that a series of stock answers, phrases, memetic retellings, and so on become ways in which the community reaffirms its culture and membership. Of course, these rituals can also become stale and frustrating to longtime community members, especially when ritualized interaction becomes something akin to an echo chamber or hivemind, wherein alternative viewpoints or ways of being are rejected in favor of yet another retelling of a favorite moment from reddit's past.

Like a traditional theatrical performance, interactions on the site also contain elements of both front stage and back stage performances (Goffman, 1959). One unique aspect of reddit is the way that "backstage" discussions of the inner workings of the site become "front stage" fodder for many of the site's meta-subreddits—and is even a regular topic in others. This preoccupation with and continual discussion of the eccentricities of the community and its denizens are in many ways unique to reddit. (I discuss this further in Chapter 5.)

Reddit as play

Viewing interactions as performances highlights how cultures are defined by and reconfirmed through ritual. For redditors, play becomes the primary mode by which reddit culture is enacted, membership is codified, and community is solidified. Early play scholarship argued against dismissing play as "mere" frivolity, emphasizing instead its connection to larger cultural practices and mores (Caillois, 2001; Huizinga, 1971). More recent scholarship highlights the slipperiness of play—the multiple ways in which it is experienced and rhetorically positioned in relation to contemporary social life and the inherent ambiguity we have in defining just what constitutes play (Sutton-Smith, 1997). Brian Sutton-Smith (1997) argues that play embodies seven different rhetorics, simultaneously being a means of power exchange, a way of reinforcing community identity, and a protest against the everyday ways of being (among others). Henricks (2006) argues that "to play fully and imaginatively is to step sideways into another reality, between the cracks of ordinary life" (p. 1). In other words, play on reddit can be said to take place inside a metaphorical "magic circle"—a liminal, permeable boundary surrounding play and gaming spaces where interactions are deemed somehow special or different from interactions in "non-play" areas of life (Salen & Zimmerman, 2003). The concept of the magic circle has been critiqued by multiple theorists for creating false distinctions between activities that are actually quite blurry, suggesting, for example, that things that happen inside play-/game-spaces have no impact on the "real world" or that play in games can be divided so easily from work (Consalvo, 2009; Moore, 2011; Pearce, 2011; Taylor, 2009). As Constance Steinkuehler (2006) argues, play in gaming spaces is ultimately an emergent collaboration between designers and players, "seemingly holding still only when the 'mangle' of designers' intentions (instantiated in the game's rules), players' goals and agency (instantiated in shared, emergent practices), and broader economic, legal, and cultural issues reach a (temporary) point of stabilization" (p. 211). As such, trying to define where play begins and ends in a space like reddit becomes challenging, as the rules which help define the space are constantly shifting and being negotiated at multiple levels by both players (redditors) and designers (administrators and moderators).

Likewise, play on reddit is multifaceted, complex, and emergent. There are, however, explicit and implicit rules that one can glean after participating in the community for some time. It is a space of emergence (Juul, 2005) rather than progression, as there are relatively few rules governing play; instead, a complex system of interactions exists that can result in the acquisition of karma—a point system by which individuals' contributions are tracked. As with traditional games, karma points have only endogenous meaning (Costikyan, 2002) for community members, and yet redditors often engage in reflexive talk about them (e.g., noting that their most upvoted comment is a story about their cat). An account with a higher amount of karma also gains a kind of reddit notoriety, whereby later comments and posts acquire more attention and karma points, creating a kind of power law effect (Shirky, 2003). Frequent, off-topic posting is not always rewarded, but moments of clever Dadaist play and non sequiturs are often upvoted and commented upon. The rules of play in this space are taken seriously while also shrugged off as inconsequential by the community. This is evidenced by almost daily discussions around what constitutes "karmawhoring" (reposting content or commentary solely for upvotes, which I discuss in Chapter 5) and the implication that somehow this is "against the rules" of reddit.

Reddit as community

Since the advent of social networking sites (SNSs) in the mid- to late 2000s, new media scholarship has been awash with discussions of how these spaces create opportunities for self-presentation and micro-celebrity while also requiring new modes of engagement given their tendency to create a kind of "context collapse" (Ellison & boyd, 2013; Marwick & boyd, 2011b). Certainly, social networking sites are popular among large audiences and deserve broad attention from scholars from a variety of disciplines and perspectives. However, we seem to have turned away from much of what characterized early new media research—the ways in which individuals commune with others in online spaces. This is not to suggest that SNSs do not facilitate or promote community, but that for many of us, the experience of SNSs foregrounds individual presentation of self.

In some ways, reddit feels more like the earliest online communities rather than being a product of the Web 2.0 era—a point also articulated by Ethan Zuckerman (2013). This is for a couple of reasons. First, like the communities discussed by early new media scholars and critics (Dibbell, 1999; Jones, 1997, 1998; Kollock & Smith, 1999; Rheingold, 2000; Turkle, 1995), reddit allows for and encourages a sense of identity that is mutable, flexible, and multiple. It is, I would argue, a repudiation of the singular model of identity championed by other platforms—most notably, many social networking services (Facebook, Google+, etc.). Likewise, the experience of reddit exists in both material and symbolic realms—on servers and through computers, at Meetups, embodied and at a distance. So, unlike early discussions of online community, which often implied that the "real world" was separate from the virtual one, my understanding of reddit is rooted in a rejection of "digital dualism" (Jurgenson, 2011).

Second, like earlier online communities, redditors communicate primarily through text to learn and share knowledge with one another. This does not mean that memes, images, and reaction GIFs are not all important parts of their vocabulary or that many redditors are not interested in interacting in this way. But if you visit and participate in the comment threads on the site, it is hard not to be impressed by the level of interest in understanding and explaining the world at large. From why a particular chemical reaction works the way it does to the relative merits of Christopher Nolan's *Batman* series, there is the sense that this might be a space in which community is at least partially created through collaborative knowledge sharing. It shares elements with Howard Rheingold's (2000) description of the WELL and its utopian vision of the digerati engaging in meaningful deliberation. Reddit is neither as nearly high-minded nor as exclusive as this, but there are echoes. Like the denizens of the WELL and other early online spaces such as LambdaMOO, many redditors are also technologically savvy and invested in geek culture and humor (Coleman, 2013; McArthur, 2009; Taylor, 2012; F. Turner, 2006).

The simple choice to create a pseudoanonymous space (instead of a wholly anonymous one or a non-anonymous one), coupled with reddit's geek sensibility, shapes interactions on the space in very particular ways. This is not to suggest that reddit is a utopian space in which only the best of humanity is on display. Instead, such identity play among redditors allows both the best and worst of internet culture to flourish.

Reddit as platform

My perspective on and approach to reddit are uniquely shaped by my fascination with and experience working in the new media industry. In particular, I am interested in the ways that technologies reflect both designer intentions and larger, unstated political-economic realities. As others have noted, platforms and their attendant algorithms, as designed objects, are inherently political

(Gillespie, 2010, 2014; Sandvig, 2013).[2] Therefore, I am heavily influenced by scholarship within science-and-technology studies (STS) and actor-network theory (ANT), both of which foreground the importance of understanding how individuals and technologies act on one another (Bijker, Hughes, & Pinch, 2012; Callon & Latour, 1981; Latour, 1988, 1992, 2005). ANT traces the connections and networks between humans and technologies to more fully illuminate the complex ways in which social and technological infrastructures are co-constitutive and productive. When examining a space like reddit, therefore, a primary concern becomes unpacking the mutual way that humans and the site's underlying technology shape interactions. My thinking in this direction is influenced by the argument that technologies function both materially and symbolically—and that focusing on one aspect in exclusion of the other does not fully reflect the complex and multitudinous ways in which technologies and our everyday lives are interwoven (Gillespie, Boczkowski, & Foot, 2014).

In focusing on the contradictions and tensions inherent in the reddit community, I am also indebted to scholarship from cultural-historical activity theory. Activity theory (AT) offers a particularly useful way of viewing human activity as mediated by tools (which are often technologies) and the tensions that often exist between our tool use and the objects and outcomes to which they are put (Engeström, 1999; Engeström, Miettinen, & Punamäki-Gitai, 1999; Kaptelinin & Nardi, 2006; Nardi, 1996). AT is particularly useful for explaining how activities are ongoing, developmental processes that embody a number of contradictions. These contradictions (tensions) are potential places for developmental growth over time (Foot & Groleau, 2011). Like actor-network theory, activity theory is both a theoretical and a methodological approach. Unlike ANT, however, AT does not assume complete equivalency between humans and technological artifacts—a subtle yet critical point. Human development and agency are foregrounded in activity theory, providing a necessary antidote to the often overly mechanistic and machine-oriented approaches to understanding our relationships to technology we often see in both academic literature and the media.

What reddit is not

Reddit is often described in a number of ways by journalists and scholars: as a "social news-sharing site," or "social sharing site," or "social news site" (Bogers & Wernersen, 2014). These different formulations suggest some ambiguity around what reddit *is* exactly (not surprisingly) and require some parsing. "Social" foregrounds the idea that reddit was (and is) centered on interactions and conversations with others, although the differences between reddit and other social sites, particularly social networking sites, are noted later. "News" or "social news" suggests that reddit is focused on news content, be it global, national, or local. In this context, "news" may come from official media outlets such as CNN or the *New York Times*, but it may also be generated by "unofficial" sources such as blogs, social networking sites, and the like. Again, the emphasis is on the social nature of the news shared—that is, news is fodder for commentary or discussion by reddit members. "Sharing" or "social sharing" emphasizes reddit's place in the Web 2.0 universe, where content may be both generated by and shared between individuals online (O'Reilly, 2005). "Site," presumably short for web site, suggests an experience inextricably tied to the aesthetics and infrastructural politics of the web.

However, all of these formulations are at once too narrow and too broad. The idea of "social sharing" is redundant—how can sharing not be "social" in some way? And the suggestion that reddit is predominately focused on news is problematic, as much of the content shared on the site consists of memes, questions posed to the community, anecdotes, pictures, and so forth. While these may be interesting or provocative or worthwhile to a niche audience, it would be difficult to categorize much of the content as "news" per se. None of these formulations really addresses a core feature of reddit—that content is both submitted/created by the community and evaluated by it using the upvoting/downvoting features. In addition, they do not emphasize that reddit actually consists of many different, smaller communities (subreddits), which have vastly different rules concerning the kind and format of content that users can submit. These descriptors also lack the idea that reddit mostly links to rather than hosts content. While subreddits, commentary about shared content, and what are called "self-posts" are hosted on reddit, the site itself functions more like an aggregator of links to external resources. Lastly, these descriptions of reddit do little to emphasize its crossover as both a web and mobile experience (mobile applications are popular ways to access reddit). They also do little to acknowledge that reddit is more of a platform or application than it is a flat web site (as suggested by the availability of reddit APIs and its availability as an open-source application).

As I alluded to earlier, reddit is not a social network. There are elements of social networking, in that redditors can "friend" others, which really only serves to highlight those usernames in a different color and provides another way of browsing site content. In addition, each reddit account comes with a rudimentary profile featuring the links submitted, comments made, and a "trophy case" where badges are displayed (such as how long the individual has been a redditor). Comments and links upvoted and downvoted can also be made public in a redditor's profile if desired. But unlike other social networking sites (SNSs) such as Facebook or Google+, the "goal" of reddit is less about a presentation of self or a public articulation of and interaction with one's social network (Ellison & boyd, 2013). Instead, reddit is focused on enabling conversations about a given link or topic among members who are likely outside one another's social networks. Reddit benefits from moving beyond the social networking model based on singular identities that other Web 2.0 sites employ, relying more heavily on a kind of "strength of weak ties" (Granovetter, 1983) to ensure that members serendipitously stumble on new content when browsing the site. And, unlike SNSs, there are no status updates—postings focus on a topic of interest (defined by the subreddit to which they are submitted). Although the discourse may diverge radically from the main subject matter, there is a general sense that interactions are inspired by the link or post. It is also not possible to see the network of individuals to whom you are connected as you might on Facebook (via friends) or Twitter (via followers). And unlike Twitter or Facebook, there is no concept of a "retweet" or a "like" tied to your account. An upvote, which might serve a similar purpose as a "like" does on Facebook, is not public—that is, you cannot click on a given link and see a list of all of the redditors who upvoted it. Reddit also differs from SNSs, which do not offer the concept of a "dislike" or whatever might constitute the reverse of the Facebook "like," by encouraging its members to downvote material they view as noncontributory to the conversation at hand. While Twitter encourages topic-based conversations through hashtags, individual contributions are necessarily short (140 characters) and mostly ephemeral. Accessing archived SNSs is easy if one is interested in an egocentric, archived experience but much more difficult if one is trying to view an entire network's activities at a given time. Because reddit is organized around links and self-posts, reactions to a breaking news story, for example, can be more easily accessed months after the fact.

Reddit is also unlike other participatory culture platforms that enable video or audio sharing, such as YouTube or SoundCloud. Like reddit, these spaces allow for discussions grounded in a particular artifact (videos and audio clips); however, the quality of discussion is relatively limited. Threaded discussions are only a recent addition to YouTube, and while individuals can give a "thumbs-up" or "thumbs-down" to a comment, these actions do little to change their visibility. In addition, neither allows "self-posts"—in which an individual shares text and possibly one or more links instead of just a link to some sort of content—unlike reddit.

Reddit is often compared to 4chan in press accounts, a series of image-based boards broken up by topic area that represent staples of geek culture such as /a (anime), /g (technology), and /v (video games) (Knuttila, 2011). 4chan is often conflated with its infamous /b/ board, a random space that has become the birthplace of many popular memes and the hacktivist group Anonymous (Coleman, 2013). It is also a haven for extreme, objectionable content and mob justice. While both sites share a geek sensibility (and it is probable that their audiences overlap significantly), several things differentiate reddit from 4chan. First, reddit's pseudoanonymity allows redditors to build a reputation over time and contact one another via private message, which is not possible on 4chan, as it is entirely anonymous. Second, 4chan is a deliberately ephemeral experience—threads are deleted after a few hours or days, and only limited external archives of the space exist. Postings on reddit are not deleted for the most part and remain accessible through the site's search engine and Google. Third, while a number of 4chan's boards are moderated to enforce content rules, /b in particular remains an "anything goes" space ("4chan Rules," 2014)—there is not the equivalent on reddit. Also unlike reddit, 4chan moderators are anonymous. This means that there is no public record of deletion of posts or banned users, nor is there any effective way to identify moderators or question their actions publically.

Reddit administrators use the term "reddit" as both a noun and a verb, meaning "a type of online community where users vote on content" and "to take part in a reddit community" ("About reddit," 2014). Thus, reddit would be much better described by the generic formulation "community platform." In terms of this book, I will refer to reddit variously as a platform, site, community, or, most generically, space. This is both for readability purposes and to highlight a particular way of thinking about reddit—as a platform to emphasize its connection to participatory culture; a site to emphasize its nature as a destination on the web; and a community to emphasize its appeal as connecting individuals.[3]

A final note: unlike both social media networks and artifact-based platforms, the commercial nature of reddit remains low key and minimal. Revenue for the site is primarily earned through ads, Reddit Gold, and RedditGifts, according to CEO Yishan Wong (yishan, 2013). This is markedly different from spaces such as YouTube, in which commerce and community exist in a delicate, often fraught balance (Vondreau & Snickars, 2010), and social networks, such as Facebook, in which users are the target of recycled "likes" and other not-so-subtle advertising mechanisms (Sandvig, 2014). This is not to suggest that reddit is "more pure" with regard to commercialism than these other platforms or that the site administrators and owners—Advance Publications—will not change this approach in the future. But I do believe that reddit is erring on the side of prioritizing community connection over commercial success right now.

Notes

1. Gawker's ongoing coverage of some of reddit's NSFW (not-safe-for-work) subreddits such as /r/creepshots highlights the unspoken relationship between reddit moderators and administrators. It suggests that the site's administrators are heavily dependent on the unpaid labor of high-level moderators, even if they are knowingly peddling unethical or borderline illegal content (Chen, 2012).
2. While I briefly describe some of the ways in which reddit's sorting algorithm works throughout this text and offer a few examples of the ways in which this might impact the kinds of material that become most visible to the average reddit user, a full explication of how it impacts discourse on the site is beyond the scope of this book.
3. Likewise, I also use the pronoun "he" to refer to generic redditors, as it most accurately represents much of the site's demographics and its overt ethos.

· 3 ·

CYNICISM, ALTRUISM, AND HATING EVERYTHING, INCLUDING OURSELVES

One of reddit's founders, Alexis Ohanian, published a book in late 2013 about his experiences with the site and its implications for internet entrepreneurism. Its title, *Without Their Permission: How the 21st Century Will Be Made, Not Managed*, is telling, as it encapsulates in a few words reddit's ethos: trust users and makers, not gatekeepers or institutions. It also captures one tension that exists (but is rarely discussed) in participatory culture: participants often embody both a kind of naïve altruism and optimism regarding the communities with which they engage but may remain intensely distrustful of larger institutions. Reddit's membership is no different—they altruistically give of their time and labor regularly to the pursuits that interest them intently and blissfully but may also remain intensely cynical of hierarchies and institutions. A would-be redditor is encouraged, as Ohanian's title suggests, to consider "making" his own world rather than having it "managed" by others. The irony, of course, is that all web platforms—even reddit—are commodified and designed spaces with their own technological logics that shape individual behavior.

These twin tendencies, toward a kind of naïve altruism on one hand and a deep cynicism on the other, do not apply just to how redditors interact with each other. For many of them, these views extend to the ways in which they view individuals, other communities, and political, cultural, and governmental institutions—and even inform the ways they see themselves. They also underscore the way in which redditors consider the complexities of what is considered "authentic" within the community.

Altruism and its limits

Scholars have often noted that online communities and participatory culture platforms function as "gift economies"—that is, systems in which individuals offer their labor for free (Barbrook, 1998). The results of this labor may be immaterial, such as a review on Zappos or helping someone diagnose a computer problem in the Apple forums, or affective, such as offering social support to a friend on Facebook or sharing a humorous meme intended to make others laugh on reddit. Some might suggest that participation in a gift economy makes no rational sense, as individuals are not receiving monetary compensation for their labor (Hardin, 1968). And yet, others note the tremendous affective and immaterial rewards that individuals often receive from contributing to and participating in these communities (Andrejevic, 2008; Baym, 2000).

Many aspects of reddit culture highlight how it operates as a gift economy. First is the foundational notion of reddit to begin with: that the community only exists if individuals are willing to submit content and vote on content that others submit. While karma points may be one way in which redditors are "paid" back for their labor, it seems unlikely that this is the sole motivation for individuals to contribute to the community. Still, for the site to exist, individuals need to be motivated enough to post links and comments.

Second, reddit's loose organizational structure requires a large army of moderators willing to create and manage content for its various subreddits. This is a significant time and resource commitment. Especially for larger subreddits, moderators must attract an audience of willing contributors, set and enforce rules around the kind of content contributors submit, manage disputes between redditors that occur publically in the subreddit's comment sections, create and modify the subreddit's Cascading Style Sheet (CSS) file as appropriate, and remove inappropriate comments and/or content. I got the sense from the moderators I talked to that the job was a bit like herding a clowder of cats. Moderators for the default subreddits are in frequent contact with the site's administrators, as these are monitored to ensure that the quality of the subreddit's submissions remains high.

A third element of reddit's gift economy is evidenced by the large number of subreddits designed to enable collective sharing and social support. Some examples include the kind of support groups around issues you might see in other mainstream online and offline forums, for example, /r/relationships, where members solicit and receive detailed advice from others about their romantic relationships, or /r/loseit, where members offer encouragement around weight-loss goals. In a more serious vein, /r/SuicideWatch provides support for individuals contemplating suicide, with moderators offering to talk privately with any individual feeling suicidal; /r/selfharm likewise offers a space for those recovering from self-harming behaviors, as /r/EatingDisorders does for those living (or supporting the recovery of others) with eating disorders. Other social-support subreddits are more narrowly tailored and reflect unique elements of reddit's audience. For example, /r/NoFap is dedicated to helping individuals ("Fapstronauts") abstain from pornography or masturbation ("fapping"). Casual visitors to reddit are unlikely to know about these niche, small subreddits as their posts rarely garner enough upvotes to show up on /r/all. The notable exceptions are the aforementioned /r/loseit and another weight-loss subreddit dedicated to pictures of weight loss, /r/progresspics, both of which occasionally appear on the front page.

The posts on many smaller subreddits, especially those explicitly designed to offer social support, suggest a real sense of community and connection between subscribers. For example, among /r/diabetes's ("Calloused fingers, caring hearts") discussions of pumps, testing strips, and how to support newly diagnosed family members, a redditor recently posted about the mental health issues that diabetes raised. In her posting, "Some days diabetes feels as much like a mental illness as a physical one," this individual noted that she often blamed herself for her blood sugars testing high, writing, "this whole pump experience is like having an eating disorder and an anxiety disorder rolled into one, just rebranded as the path to health" (anniebeeknits, 2014). Other redditors shared their own stories regarding their experience with pumps and diabetes generally, offered suggestions to the original poster (OP) about how she might find support from her friends and family, and provided their own logging strategies as examples of how they manage their illness. While reading this posting, I was struck by the kindness and thoughtfulness of the interactions between the OP and other redditors.

The term "gift economy" can also be used metaphorically to suggest that individuals offer affective or other immaterial goods in a community (Braitch, 2010). In other participatory platforms, it might include writing and sharing Harry Potter fan fiction, for example, or remixing and sharing images licensed under Creative Commons, or troubleshooting others' computers on a support forum. However, more material goods may also change hands. On reddit, subreddits like /r/Random_Acts_Of_Amazon (RAOA) or /r/Random_Acts_Of_Pizza (RAOP) (a play on the phrase "random acts of kindness") fulfill this function. On RAOA, redditors post their Amazon Wish Lists so others can purchase things from Amazon on their behalf. Likewise, RAOP is dedicated to people asking for and those willing to provide pizza for other redditors. Often the reason for requesting pizza assistance involves an inability to pay for that night's dinner (lost jobs, disabilities, etc.); other requests state that the individual is depressed and lonely and hopes that receiving pizza will help him feel better. Still others ask for pizza simply because they are craving one. Pizza requests that are most likely to be fulfilled are those that suggest the need is an urgent one, offer to "pay it forward," are requested by well-respected members (as evidenced by positive karma and/or contributions to RAOP community), and demonstrate gratitude (Althoff, Danescu-Niculescu-Mizil, & Jurafsky, 2014).

Other subreddits provide a wide range of material assistance to individuals both within and outside the reddit community. For example, /r/Assistance provides a space for redditors to request financial assistance for medical bills, unexpected expenses, utilities, and rent or job assistance, food, or social support in the form of cards or notes. The subreddits and the community at large also mobilize around current events. For example, during the investigation of the Boston Marathon Bombing, redditors organized to provide pizza for the precincts working the case (workman161, 2013). After the 2010 Haitian earthquake, redditors raised over $185,000 USD for Direct Relief International (kn0thing, 2010).

Beyond these explicit community efforts, individuals or small groups of redditors often provide ad hoc support for others expressing a need. This may be anything from a toy shopping spree for a young girl diagnosed with Huntington's disease who was being bullied (hmasing, 2010) to helping a man who worked at a Kenyan orphanage build a fence after he was attacked by someone wielding a machete (TheLake, 2012). Even individuals who do not request assistance but are perceived as somehow deserving of it anyway may receive donations from reddit users. For example, Darien Long, a security guard and manager working at an Atlanta mall, received almost $24,000 USD from redditors after a video of his was posted to /r/JusticePorn in which he was physically threatened by several individuals, including two mothers, one of whom he later Tasered (Mylaptopisburningme, 2013). The mall in which Long worked was known for being a market for counterfeit goods and attracting local drug dealers selling their wares. As a result, Long regularly videoed his interactions with mall patrons using a chest-mounted Go-Pro camera and uploaded them to YouTube (Mathis, 2013).

For many redditors, Long became a folk hero because of his actions. Of course, the reality was not so clear cut. In the weeks after the video was posted, Darien Long was fired from his job for being too aggressive with patrons and after being arrested by the Atlanta Police Department for allegedly tackling someone without cause (Carmichael, 2013). As Rodney Carmichael (2013) argued in an article for independent newspaper *Creative Loafing*, even the mall itself represented a much more complex story of racial and class politics and how revitalization efforts in downtown urban areas often fail to include those who are most impacted by the changes that are wrought. The Darien Long case also highlights the tendency of the reddit community to "judge first" and ask questions later—a problematic approach given the complex realities of life outside of reddit.

RedditGifts

There are also other more formalized material efforts that highlight reddit's nature as a gift economy. Individual subreddits will often host gift exchanges based on certain themes. For example, /r/MakeupAddiction has hosted several gifting events around makeup and beauty products for its subscribers. Likewise, /r/books and /r/movies have hosted similar exchanges in the past. While smaller subreddits may coordinate these kinds of events on their own, larger subreddits very often rely on redditgifts.com, which hosts exchanges throughout the year. These may include themed exchanges around specific popular culture franchises (such as *Dr. Who* or *Harry Potter* or *Lord of the Rings*), objects (such as lunchboxes, T-shirts, socks, or coffee mugs), edibles (condiments or snacks), or holidays. The most famous exchange is the yearly online Secret Santa that takes place in December and has repeatedly broken the record for the largest online gift exchange by the *Guinness Book of World Records* ("Nine Things to Know about RedditGifts," 2014). The 2013 exchange involved 121,879 individuals from 189 countries (highshelfofsteam, 2014). Celebrities such as Bill Gates, Arnold Schwarzenegger, Shaquille O'Neill, and Wil Wheaton, among others, have participated in reddit exchanges. The popularity of the Secret Santa exchange has also spawned another large gift exchange based on a reddit-created holiday celebrating the halfway point between Christmases (June 25) called "Arbitrary Day" ("Arbitrary Day 2014," 2014).

To participate in exchanges, a redditor must spend what are termed "exchange credits." You receive an exchange credit automatically upon signing up for RedditGifts, which means you can immediately participate in an exchange. You also receive an exchange credit after your giftee indicates having received a gift. This system presumably keeps people from gaming the gifting experience by simply signing up for an exchange as a receiver and never sending a gift in return, using the same reddit account over and over. However, it does not prevent individuals from creating and signing up with a new account each time and simply not completing the exchange. On matching day, Santas receive the username and address information for their giftees. Santas are encouraged to "stalk" their giftee, but not in a "creepy" way—meaning they can examine their giftees' reddit profiles or search for them on Google, Facebook, and the like to generate gift ideas but are on their honor not to send inappropriate private messages or otherwise use contact information for anything other than the gift exchange ("RedditGifts FAQ," 2014). After sending the gift(s), the Santa marks them as shipped. Once the giftee has received the gifts, he/she is required to confirm receipt by posting to the RedditGifts gallery.[1] If a Santa fails to send a giftee something, the giftee may be rematched after a period of time with someone else who is willing to give another gift. Santas are not required to reveal their identity to their giftees unless they so choose. The guidelines suggest that Santas should spend about $20 USD on their gifts, but frequently they spend significantly more. Santas have sent giftees a cruise (askusmar, 2011), Apple iPads (Johnmtl, 2010; lacylola, 2011), and computers (runningman_23, 2011; sleepytotoro, 2012). Creativity in terms of the gifts and the packaging of the gifts is encouraged. For example, a redditor sent another on a "social scavenger hunt" to help him to overcome social anxiety (Hilarious_Exception, 2013); another sent a giftee a four-part choral piece based on a Pablo Neruda poem he/she had written (ladyofchaos, 2011); and one redditor received gifts packaged inside a plush shark that required "operating" on the shark to retrieve them (Believeinfacts, 2009). Some gifts invoke reddit's more playful and bizarre humor, such as the 22-pound box of Idaho potatoes one giftee received (thanks_for_breakfast, 2012) and the ironic gift of a sex toy and lubricant a moderator of /r/NoFap received.

In December 2013, I took part in reddit's Secret Santa exchange. This involved indicating my likes, dislikes, favorite subreddits, and T-shirt size and noting my favorite television shows, movies, and books. I also wrote a letter to my Santa suggesting some ideas for what she/he might gift me. My note was rather lengthy and included information about my profession (college professor), what I taught (games, new media, etc.), and the things that I like (DIY, design, crafty stuff). I also provided links to my Amazon Wish List and Kickstarter profile, as I thought they might help give my Santa a more specific set of gift ideas. Because it was my first exchange, I decided to sign up as a RedditGifts "Elf." Individuals become Elves by making a donation to support the RedditGifts project or are gifted Elf status because of the high quality of the gifts they have given in the past. In return, Elves can be matched with other Elves for certain exchanges—thus making it more likely that an exchange will actually happen—and receive a special "Elf" trophy for their reddit profile page. During the sign-up process, I also indicated my willingness to ship abroad and offered to be a rematcher.

I was matched to an Elf living in Norway who provided a list of interests. After doing a bit of sleuthing on reddit to find out where he posted regularly, I compiled a list of potential gifts for him (and, this being reddit, his cat). Unfortunately, I had to rely on Amazon UK, as shipping from the US to Norway is staggeringly expensive. And, for some reason, Norway has tight import restrictions, so several of my original gift ideas (catnip toys for the cat, coffee for the redditor) had to be rethought. I settled on sending him a couple of books and a card game. After seeing all of the creative gifts others sent, I felt a bit underwhelmed by mine, especially because I could not personalize the packaging or wrapping. But it was gratifying to give a stranger something that I thought he'd enjoy, even though I would have liked it to have been a bit more personal.

My Santa, on the other hand, was awesome. He/She was also overseas, stationed on some sort of ship, and had to rely on my Amazon Wish List for gifts. That being said, my Santa went way overboard, sending me several books, a video game, the complete *Daria* series, and a number of T-shirts featuring vintage, minimalist graphics depicting famous scientific inventions that I had been eyeing for some time. I was actually astonished by my Santa's generosity—and told him/her so on several occasions. My experience mirrors that of other individuals who have also participated in reddit exchanges. One of my informants, Carter, suggested that the overall altruism of the reddit community was admirable, writing:

> I love how strangers reach out to one another here. I love participating in reddit gifts and I do poke around charity subs like food pantry. I think it is neat that strangers on the internet can have such a huge emotional and/or physical (be it financial or otherwise) impact on one another.

When I asked this individual about her experiences with RedditGifts specifically, she mentioned that she was most interested in giving to others rather than receiving anything in return: "it is all about the giving. I have been shafted in some exchanges but I don't really care about it. I was more upset when it looked like my gift didn't make it there." For this person, connection with others who are "strangers" outside of reddit is valuable. And getting feedback from giftees is particularly important, as one of my informants noted. She said that the only disappointing thing about the gift exchange process was not getting feedback from someone whom she had sent gifts to, as this was a significant part of what motivated her to participate. For this person, it was less about the gifts that she received and more about feeling connected to other redditors and reveling in the pleasure of giving generously: "I think even if the gifts I get are somewhat disappointing, the reactions I get from my giftees are worth it. I spent 70–80$ on the snack exchange and my giftee was super psyched when he got it. I participate for the good feeling I get."

Another informant, Armani, said that part of what made RedditGifts so special was the fact that most people participated in good faith:

> I have participated in Redditgifts. I REALLY like the idea of redditgifts. I'm pleasantly surprised by how many people aren't cheating the system. People are completely on the honor system here—I can easily lie and say I never received a gift to get a second one. I can easily lie and say I sent a gift, and no one would question me much further. However, from what I've seen, the redditgifts thing works great … people are more than happy to send and receive gifts from random strangers and no one is really held accountable for their actions. I'm sure that there some people cheating, but for the most part, everything works. It makes me think: "Maybe there's hope for us yet, and maybe the world isn't completely depraved."

What I find particularly notable about this individual's response is that it mirrors both sides of reddit's altruism/cynicism dichotomy. There is the idea that people choose not to cheat the RedditGifts system despite the relative ease of doing so, instead participating in good faith, and yet there is the sort of classic reddit cynicism regarding the general view of humanity as "depraved" and mostly hopeless. For the reddit community, many of whom likely spend their days behind screens, as STEM college students or working in IT, the material and physical nature of the RedditGifts experience is probably especially enjoyable. Given the creativity on display in the RedditGifts gallery with regard to both the appropriateness of the gifts and the detail spent on packaging them uniquely, it is clear that many redditors enjoy the potential of connecting with other community members in some physical way. Perhaps, too, the anonymous nature of the various gift exchanges appeals to redditors' desire for connection with their online counterparts and an interest in maintaining their own boundaries regarding privacy.

Comment gilding and tipping

Redditors often show their appreciation for others within the community in other ways. Individuals can purchase what is called "Reddit Gold" for themselves or others.[2] Reddit Gold serves mainly as a way for redditors to support the site's server costs and yields several minor benefits: access to a special subreddit (/r/lounge) and a reddit gold trophy for your profile page, the ability to show comments that have appeared since you last visited a posting, notifications if your username is mentioned in the comments, ability to filter certain subreddits from appearing on /r/all, an option to turn off advertisements, and discounts at certain reddit-friendly retailers (yishan, 2014b). Individuals can either give another redditor gold for his/her account as a whole or gild a particular comment they found particularly insightful, funny, interesting, or useful. Oftentimes, celebrities who visit the site regularly (such as William Shatner) or are particularly beloved by the reddit community (Bill Gates) or both (such as *Scrubs* actor Zach Braff) may be given gold for their contributions. This usually results in at least some comments being made by other redditors questioning the reason these individuals received Reddit Gold, as they may not be regular site visitors or particularly invested in the community. Even more confusing or amusing, depending on your perspective, is that bots (automated scripts) often receive gold from other redditors. (I discuss bots in Chapter 5.)

Individuals may also gift and receive tips in the form of Bitcoin (a cryptocurrency) or Dogecoin (a Bitcoin offshoot).[3] These are given out at random by other redditors using a bot—in the case of Dogecoin, /u/dogetipbot. As with Reddit Gold, individuals may be tipped for an incisive or useful comment or at the whim of the tipper. The denizens of /r/dogecoin in particular, the subreddit dedicated to all things Dogecoin, are known for their generous tipping, especially to currency newcomers. Dogecoin is derived from a well-known meme, called "doge," which originated from a series of pictures of a Shiba Inu dog sitting daintily on a couch, giving its owner a knowing look. The archetypal doge picture features lowercase pidgin English phrases (such as wow, so doge, very excite) written over the photo in multicolored Comic Sans (NovaXP, 2013).

Cynicism on reddit: the case of /r/cringe and /r/cringepics

While there are moments of altruism and interactions that suggest an almost naïve willingness for redditors to trust other redditors, individuals also express deep cynicism toward both institutions and others. This tendency often manifests itself in some of the site's more fringe (but still popular) subreddits. For example, both /r/cringe and /r/cringepics host content that makes the viewer "cringe" in embarrassment for the participants. /r/cringe traffics in videos "that are too embarrassing to watch all the way through" as the sidebar claims. I found characterizing exactly what was "cringeworthy" enough to make it to the top of /r/cringe (and its picture-/screen-capture-related sibling, /r/cringepics) difficult, mostly because viewing the comments on these images became emotionally draining. But it becomes apparent that there are several categories of "cringe." Many top-rated /r/cringe videos involve teenagers doing what teens often do—pushing social boundaries, joking around with friends and classmates, and performing privately or publically. But the linked videos evoked a familiar sense of dread and embarrassment in me as I viewed them. There are the young woman performing a ventriloquist act at a talent show that goes south as she forgets her act; the young man hoping to inspire others before a test with a speech from *Lord of the Rings* while his classmates simply stare; and the college student impersonating the Joker when his roommates walk in unexpectedly.

These videos remind me of my childhood, and I assume this is probably true for many redditors. I was a theater kid/teen and can remember screwing up the courage to get on stage and perform only to have it go horribly, horribly wrong. I can vividly remember times when I wished the floor would open up and swallow me whole so I could avoid the crushing embarrassment of what would be awaiting me backstage and the next day at school. Moments like these are reminiscent of a scene from the film *About a Boy*, in which a young boy (Marcus, played by Nicholas Hoult) sings Roberta Flack's "Killing Me Softly" a cappella at a talent show. Marcus's performance is so earnest and guileless that the ensuing fallout not only demonstrates the cruelty of kids but also makes the viewer wonder why nobody stopped him before he went on stage. I can only imagine that others feel the same way. As one redditor notes in a highly upvoted comment about the female ventriloquist's performance,

> I always re-cringe myself after videos like these when I think about the process she went through leading up to this moment. Imagine how she mulled around the idea of

doing the talent show before finally deciding in her own mind it was a good idea ... She maybe rehearsed it with her little buddy in the bathroom mirror a couple times before showing the routine to her mom and dad. Of course they laughed and laughed and gave her the encouragement she needed ... She lied awake in her bed visualizing how awesome the performance would be. She saw people in the audience laughing heartily while her friends were overwhelmed with jealousy ... Little did she know thousands of people would later see it on YouTube and Reddit. *cringe*. (antizeitgeist, 2012)

antizeitgeist's comment suggests that cringeworthy events are likely the result of naïve intentions on all fronts—from the "performer" captured in these videos to those surrounding him/her who wish to remain encouraging. At another level, the viewer watching the cringeworthy moment may feel a deep sense of empathy for the person(s) in the video or picture but is likely also to feel gratitude that his/her own mishaps have not been captured and shared in such a public way. But this empathy is also complicated by the fact that these videos are uploaded to public sites such as YouTube, either by the person performing or someone in the audience, and then shared on reddit. Thus, the cringeworthy event takes on a life of its own, as it is replayed, commented upon, and archived. And as the goal of /r/cringe is ultimately entertainment—it is not solely about encouraging empathy in the "audience" or supporting the "performers"—it remains space for guiltless *schadenfreude*.

Top posts in the /r/cringe and /r/cringepics subreddit include several categories. The first, as mentioned above, contains videos of teens/preteens. Despite the fact that videos with minors are no longer allowed, it is telling that the all-time top submissions for the most part feature teens. A second category of top-ranked videos involves humiliation of public figures. For example, there are clips from reality television shows in which characters seem oblivious to the effect of their actions on others, newscast bloopers, and celebrities behaving awkwardly (for example, Tyra Banks acting as though she has rabies on her talk show) or inappropriately given the seriousness of the circumstances (for example, Justin Bieber's criminal deposition in 2014). Third, there are the videos and pictures of adults displaying a kind of awkwardness we typically associate with teenagers—for example, individuals dressing and/or acting "inappropriately" given their age. On /r/cringepics, these postings often include screen captures of exchanges that occurred on Facebook or over SMS messaging or pictures of tattoos that are misspelled or offensive or profile pictures downloaded from the OKCupid dating web site.

I found some videos and pictures posted to the two subreddits disturbing less because of the content they contained but more because of how they are received by subscribers. For example, there is the NSFW (not-safe-for-work) video of a preteen boy "twerking" in his school uniform to Rhianna's "Birthday Cake." He is a damn good dancer, and also effeminate, which predictably means that many of the most upvoted comments are remarks about his presumed homosexuality or outright homophobic slurs ("Kid Attempts to Twerk in His Private School Uniform to 'Birthday Cake,'" 2012). Interestingly, one commenter notes that it's not the video itself that is "cringeworthy" but the act of watching the video that constitutes "cringe"—"The way he acts (flamboyantly gay) isn't cringeworthy ... Recording yourself performing a common dance isn't really cringeworthy either ... What's cringeworthy is the *act* of watching the video. Being conscious of yourself while watching the video is what causes so much embarrassment. Like I'm hoping nobody ever finds out this was on my computer" (robwinnfield, 2012). This redditor echoes a comment many others make in the thread: that a primary "cringe" factor is the reaction that others might have if they were observed watching the video.

In the case of the young boy twerking, there seems to be a general consensus that it was not clear how the subject of the video might be viewed by others. But other cringe postings focus on the perception that the person *should* "know better" about how he/she is perceived by the world at large, as judged by reddit, of course. For example, one highly upvoted posting to /r/cringepics that made it to /r/all featured a selfie picturing a white female Juggalo (fan of the band *Insane Clown Posse*) with her face painted in clown makeup, her bright red hair pulled back in two ponytails, wearing a short strapless zebra top with her bra straps visible (moab-girl, 2014). Presumably, the main reason the image was submitted to /r/cringepics and considered "cringeworthy" enough to be upvoted by other redditors is that the woman is overweight, outrageously dressed, and looking much older than her 23 years. One top-voted comment suggests that she dresses like this solely for the attention and implies that she actually lacks any personality at all; another writes, "she looks like she smells terrible." More important, her status as a Juggalo marks her for many redditors as being lower class and "trashy," a point that many of the comments imply or state outright. However, another set of redditors suggest that the insults constituting the most upvoted comments in the thread amount to bullying, with some people suggesting that reddit would respond differently if the woman were 20 pounds lighter and others arguing that insulting someone for fashion/music choices or desire to post selfies is mean, or as one redditor notes, "Honestly, who is she hurting? But I think we're in the wrong thread, mocking and jeering is the raison d'etre around here it seems." Unfortunately, these statements linger at the bottom of the thread as they have few upvotes, so the top-ranked comments, which are almost universally insults, remain the ones most redditors are likely to see.

What is more troubling about exchanges like this is that they are located underneath an anti-bullying statement that appears on every page of /r/cringe and /r/cringepics: "Keep your comments civil, bullying will not be tolerated. Report any comments that are bullying in nature." What constitutes the bullying behavior in this case? Is it the original poster (OP) who chose to share this picture of someone from another social networking site that he knows in some capacity, given his assertion that "she's 23 and goes out like this in public"? Or is it the prejudicial comments and insults that other redditors hurl at the person pictured or videoed that still other redditors upvote? Given reddit's strong pro-privacy stance, couldn't the very fact that a picture of a person was shared without her consent be considered a privacy violation, as it would not be difficult for this young woman's family, friends, or potential employers to identify her? This is not simply a hypothetical situation, as a recent posting to /r/SubredditDrama (a subreddit I discuss in the next chapter) suggested that a person had come across a picture of his/her brother on /r/cringepics that was likely uploaded by an ex (neuroticfish, 2014). The fact that few comments on the postings I analyzed are actually marked (deleted) (which denotes postings that either the commenter or a moderator has deleted) suggests that neither the subreddit's members nor the subreddit's moderators consider these kinds of interactions bullying. Perhaps it is because offensive comments on the video or photo remain in the silo of reddit instead of being linked on the individual's social networking profile on Facebook or Twitter—even if the realities of one's "brand" in the Web 2.0+ era do not remain so neatly compartmentalized (Marwick, 2013).

In addition to the anti-bullying banner, a number of all-time top posts to /r/cringe and /r/cringepics feature discussions reminding redditors to refrain from bullying or posting inappropriate content and seem to dominate many of the top postings, suggesting that it is an ongoing issue. These include rejoinders not to post "kill yourself" or other offensive comments on YouTube videos featured on /r/cringe (judgementbarandgrill, 2012) and not to submit material featuring disabled individuals (Al_Simmons, 2013). Postings like these receive a variety of responses. One category of comments justifies the enjoyment of /r/cringe because "Cringe isn't laughing at people, it's feeling embarrassed for them. It's not as bad." Still others suggest that videos uploaded to spaces like YouTube are public, and therefore the uploader should expect that he/she might be on the receiving end of negative comments. As one redditor writes, "So if I decide to upload a video not expecting a hate-brigade, I can't opt out once I start to get bullied?" to which another responds, "This isn't how the Internet works. There is no edit-undo button. You post it, you deal with it ..." More disturbing are comments that suggest bullying is only inappropriate because it results in less fodder for subscribers. As one redditor argues, "What's even worse is that those negative comments make people remove their videos, which means other cringers can't enjoy them," with another agreeing in response, adding, "This, mostly this. It is our responsibility to preserve these cringes for all."

At the same time, active /r/cringe redditors regularly post or comment about the perceived lack of "good" content in the subreddit. For example, a posting in late 2012 by Dao_of_Tao offered a rejoinder to the community, suggesting that "we've been losing sight of what *is* cringe-worthy vs. what's just kind of embarrassing and humorous" (Dao_of_Tao, 2012). The OP offers an example of what constitutes cringe: reality star "The Situation" from *Jersey Shore* presenting at Donald Trump's roast.[4] Then the OP links to three more "not cringeworthy" examples—one, a video of a song sung by Christian kids about atheists titled, "My Faith,";[5] another, a faked video of pop star Enrique Iglesias singing off-key;[6] and a web site, the "Tween Jesus & Me" category of Christian media company Manaka Bros.[7] Dao_of_Tao suggests these aren't cringeworthy as they "aren't painful to watch, they aren't really awkward and they aren't embarrassing." Comments to the thread vary, with some redditors agreeing that /r/cringe content was declining in quality (with some calling for the moderators to be more proactive in removing posts) and others decrying the fact that /r/cringe is filling up with postings talking about how crappy /r/cringe is becoming. These kinds of exchanges suggest a kind of circular logic when juxtaposed next to anti-bullying statements. If /r/cringe must continually improve (or, at least, not degrade), the community must seek out and post even more outrageous and little-seen content. This could increase the likelihood of material that was never intended for public dissemination or was uploaded in good faith that it will not be circulated.

Of course, /r/cringe and /r/cringepics are just two of reddit's many communities that share images, screenshots, or videos of individuals deemed worthy of reddit scorn for one reason or another. There are, in my mind, the even more objectionable fatphobic subreddits like /r/fatlogic and /r/fatpeoplehate, which traffic in images and stories serving to shame fat people. While the former limits "fat hate" and bullying (according to its sidebar), the latter is an anything-goes space filled with images with titles such as "NICE TRY FATTY," for an image of a woman that has been altered to make her look thinner, and lines such as "Yeah like she's ever used that equipment," which is the title for a posting featuring an overweight woman posing on a weight bench. It is hard not to view this kind of material as both unethical and potentially libelous, although, of course, no obvious personal information about the individuals, such as their names, locations, or employers, is shared. Without question, the discourse on /r/fatpeoplehate verges on bullying and further entrenches a critique of reddit's tendency toward misogyny, as almost all of the images feature women, which I discuss further in Chapter 6.

Activism and mobilization

A number of specific moments of political mobilization across reddit suggest a thread of common values that many site members hold. Not surprisingly, much of the activism surrounds protecting free speech and online infrastructures within the US, most notably issues of net neutrality, piracy, and privacy. The Stop Online Piracy Act (SOPA) and its related proposed bill Protect IP Act (PIPA) were attempts by Congress to require ISPs to engage in more tracking around internet use in an effort to stem online piracy; both became a focus of reddit mobilization (Dachis, 2012; Dixon-Thayer, 2014; "SOPA and PIPA Bills Lose Support," 2012). Perhaps engagement with these issues is merely self-protective, as some subreddits skirt the line between legal and illegal speech (sharing content that would be considered unethical at the very least). One gets the sense that the interest redditors have is also self-serving in other ways, as many of them are connected to the high-technology industry and would be adversely affected by these kinds of regulations. And certainly the founders' commitment to open-source technology might create some pressure for the reddit community to be active around issues of free information access (Coleman, 2013). That being said, there is a sense among my informants that only a small portion of the community is actually interested in these kinds of political issues, while others may be merely engaging in a kind of armchair, low-effort "slacktivism." Still, as Christensen (2011) argues, viewing these kinds of behaviors as mere "slacktivism" undermines the possibility that these activities may contribute to an individual's overall feeling of political efficacy while also underestimating the potential for such behaviors to encourage offline action. Regardless of whether or not political activism on reddit around net neutrality and other open-technology initiatives is actually effective, it is hard to fathom that given the prevalence and visibility of reddit's efforts on these issues, that redditors visiting the site during these moments do not become more aware and more informed about them.[8]

However, the reddit community is not wholly progressive in its attitudes. For example, when one administrator affiliated with RedditGifts encouraged the community to support marriage equality efforts in Utah (where RedditGifts is based), many upvoted comments were shockingly regressive and/or decried the concept of reddit exerting its political will in any form (maxgprime, 2014). This is despite relatively little controversy over reddit's involvement with SOPA/PIPA/net neutrality issues. A number of comments demonstrated a limited understanding of what the First Amendment actually protects (not the rights of redditors, as reddit is a private entity), and still others offered long diatribes about the apparent unfairness of the term "marriage equality" as "equality" would campaign for arrangements such as polygamy as well. So the idea that redditors on the whole will support any issue that site administrators do is not true, nor do they offer blanket support for progressive causes. Instead, one gets the sense that reddit as a community holds deeply libertarian views—pro-drug legalization, anti–big government, pro-business, but anti-interventionist. This is, after all, a space where the animated GIF of Ron Paul waving his hands with the words "It's happening" (z0mbi3jesuz, 2012) is often used in /r/circlejerk and other meta-subreddits to indicate reddit's preoccupation with the politician.

Several other forms of activism on reddit are less political but point to an important function of the community using mobilization as a form of play (a point I explore in depth in Chapter 5). One example was the crowning of "Mister Splashy Pants," a humpback whale that reddit was integral in naming. In 2007, Greenpeace solicited name ideas from the public through an online poll to raise awareness about the threat that the Japanese fishing industry posed to whales. Reddit's formidable power in mobilizing members toward some action, particularly, it seems, if it is either in line with the sensibilities of its members or plainly absurd or funny, led to the satellite-tracked humpback being named "Mister Splashy Pants" (_black, 2007; Nicole, 2007; TEDIndia, 2009). Communities mobilizing to "game" certain online polls or crowdsourcing efforts are not a new phenomenon or one limited solely to reddit. The site 4chan is known for efforts to manipulate votes so that Justin Bieber would play a concert in North Korea (SteveTR, 2011). However, it is typically easier to trace the direct efforts of reddit's impact on such events as posts are archived and members make reference to them long after the specific event has passed (see Chapter 4 for more about reddit's preoccupation with its own history).

Reddit detectives or reddit mob?

The flip side of the potentially positive, or at the very least relatively benign, role that reddit plays in mobilizing around social issues is that interactions can often lapse into a kind of mob justice that can have dire consequences for those targeted. My informants also noted this dual nature of reddit—that the community displays impressive largesse, especially when it comes to individual needs or collective political efforts, but can easily engage in mob-like behavior when "crossed." As interviewee Carter noted, "Redditors can really do wonderful things … There is a tendency to move the Earth and Sun to do the 'right thing' as determined by the hive mind. That same tendency can do ugly things too though. When the pitchforks come out, there can be some seriously awful consequences." What constitutes being "crossed" seems up to debate, but it can include actions such as Gawker's unmasking of popular but controversial moderator Violentacrez, which eventually led to a number of subreddits banning submissions that pointed to any of Gawker's collection of media properties (Gawker, ValleyWag, LifeHacker, Jezebel, etc.). Or it may simply be perceived as being inauthentic or misleading (as I discuss in the next section).

Like their more chaotic 4chan cousins, the immense power of reddit's collective intelligence (Jenkins, 2006) and the enjoyment some redditors seem to derive from seeking out and sharing hard-to-find information sometimes result in unsubstantiated rumors being shared as if they are empirical facts. One of the more extreme cases of this occurred in the aftermath of the Boston Marathon Bombing in April 2013, in which redditors were quick to mobilize—creating a subreddit specifically for sharing information about the suspects (/r/findbostonbombers). While some posts on other subreddits were dedicated to getting help to those affected by the events (such as /r/news and /r/boston), posts in /r/findbostonbombers were focused on determining who might be responsible. Redditors on the ground in Boston as well as those at a distance posted and analyzed photographs of the crime scene, transcribed and deciphered dispatches from local police scanners, and engaged in an ad hoc and yet coordinated effort to determine who might be responsible. This kind of reddit outlaw "mob justice" was high profile enough, and some theories seemed plausible enough, to pique the interest of law enforcement officials in Boston (who, of course, also may have been redditors themselves). Watching this unfold in real time was both frightening and fascinating. Using security footage taken during the marathon, many of the posts pointed out specific individuals in the crowd. Some redditors offered theories about this or that person who seemed to have a backpack in one image but not in another, while others chimed in that this provided incontrovertible proof that said individuals were clearly perpetrators. Later, other images would surface, and another theory would be posited suggesting a new version of what happened and who might be responsible. For all of reddit's posturing as a community that champions skeptical and rational discourse, the tone in the immediate aftermath of the bombings was one of both understandable outrage and a misdirected effort to "help" or outdo law enforcement's investigation efforts.

Eventually, the denizens of /r/findbostonbombers became convinced that Brown University student Sunil Tripathi was responsible for the bombing. Missing for several weeks at the time of the events, reddit's speculation that Tripathi was involved was convincing enough for local news media to stake out the family's home (Bidgood, 2013) despite there being no concrete evidence tying him to the events. A few days later, Tripathi's family announced that his body had been found in a river near Providence—his death resulting from unknown causes. As Alex Hern of the *New Statesman* argued, reddit's "initial hunt to find the bombers devolved rapidly into a sort of 'racist Where's Wally,' profiling—racially and otherwise—scores of innocent people" (Hern, 2013). In an interview with *The Atlantic Wire*, /r/findbostonbombers founder (and aptly named) "oops777," admitted that the entire endeavor was "naïve." In regard to the sharing of private information on the subreddit, oops777 said,

> Overall, it was a disaster. It was doomed from the start when you look at it in hindsight, because not one of the images that were available on the Internet actually had the bombers in it. I also fully admit that I was naive to think that everyone would listen to the rules and keep the posts within the subreddit. (Abad-Santos, 2013)

At the same time, oops777 argued that reddit should not be held to the same level of accountability that other news media are or used as a source, since "It's no different from a newspaper printing 'a guy on the street said, "My mate told me that this guy is a bomber"'" (Abad-Santos, 2013). Still, part of the appeal of reddit is the ease with which individuals can harness the collective intelligence of the community, but the lack of oversight means little accountability when things spiral out of control.

These events point to the potentially darker purposes to which collective intelligence can be put, and they also underscore two tendencies within the reddit community. The first is the tendency for redditors to enjoy (and be rewarded for) sharing obscure or discipline-specific information with others. This desire may trump other concerns, such as ethical considerations. Instead of the mundane, plodding police work that was happening on the ground in Boston following the bombing, /r/findbostonbombers was filled with wild speculation driven not only by the technological prowess of its contributors, especially when it came to tracking down personal information about individuals online, but also by a desire to solve this particular puzzle and "best" the police.

The second tendency that the Boston Marathon Bombing debacle points to is redditors' underlying distrust of law enforcement and the expanding US surveillance state since 9/11. Of course, there is an ironic tension here—the /r/findbostonbombers contributors spent much of their time analyzing security camera footage to determine who might have placed the bombs, even though the reddit ethos is one of deep suspicion of governmental surveillance. Redditors tend to espouse cyberlibertarian ideals (Borsook, 2001; Kelemen & Smith, 2001; F. Turner, 2006; Winner, 1997), which likely reflect a larger, self-reported trend for members to be both tech-savvy and young white men (Duggan & Smith, 2013). Thus, reddit's response to the Boston Bombing brings the tension between altruism and cynicism to the fore. The intense desire to be of service—by collectively sorting through, analyzing, and perhaps discarding a large amount of evidence in the aftermath of the event in an effort to help/one-up law enforcement—was perhaps driven by a cynical view of law enforcement's ability to do this effectively. And trumping the desire to keep private information off the site was overridden by a desire to find the person who committed the bombings.

It is also important to note that reddit was not the sole purveyor of speculative information after the bombing. The *New York Post*, for example, ran a front-page picture based on reddit's intel of two individuals it believed were involved (Popper, 2013). Later, the two men sued for libel (Haughney, 2013). And CNN suggested inaccurately that arrests had been made early on in the bombing's aftermath, a point repeated by several other news outlets including the Associated Press (Fung, 2013). But the difference between reddit and these organizations is that the *New York Post*, CNN, and the Associated Press are media outlets that employ a staff of journalists, editors, and ombudspersons to oversee the production of news content and address mistakes when they are made. Reddit is merely a loose connection of individuals with no formal journalism training or official process for addressing mistakes and grievances. However, the speed and intensity with which redditors would sift through mountains of data around the bombings made their findings tantalizing to journalists, who in the age of decreasing budgets for investigative journalism now rely on the work of bloggers and others to supplement their own research (Bivens, 2008; Bruns & Highfield, 2012; Tremayne, 2007).

After the event, redditors' penchant for meta-commentary (explored further in the next chapter) extended into lengthy discussions of how unsuccessful /r/findbostonbombers' efforts were and how much negative media criticism the site as a whole received because of the debacle that ensued after the event. At moments like these, it feels as though the reddit community members almost take their failings as a point of pride: ashamed, yes, that some members botched things up so badly and that real lives were impacted but also secretly thrilled that the mainstream media finally noticed their collective ability to mobilize in unprecedented ways. At the same time, the failure of reddit to provide any meaningful evidence in the case has become a punch line for some members. For example, in a recent /r/chicago posting regarding an Amber Alert (an emergency alert issued when a child is abducted), a commenter noted that the child was found unharmed in the suspect's vehicle. Another commenter replied, "We did it! Like the time we found the boston bombers." To which another individual added, "We're 2 for 2! Wait …" And, yet another offered a further revision, "1.5/2" (Brandtlyc, 2014).

Pseudoanonymity, authenticity, and the reddit community

Reddit prizes and expects a high level of "authenticity" from those who post to the site. But as with all things reddit, what constitutes "authenticity" is a multifaceted, complex, and often contradictory concept. Unlike social networking and other Web 2.0/2.0+ sites, authenticity is not equated with a unitary identity or a "real name" policy (Hogan, 2013). The importance of pseudoanonymity on reddit cannot be overstated; I believe it fundamentally connects to the reason certain behaviors in this particular space are encouraged. The pseudoanonymous space encourages, I believe, reddit's tendency toward both altruism and cynicism (as well as its encouragement of play as a mode of interaction—a point I discuss in Chapter 5).

Creating an account on reddit is easy and requires very little personal information—at most an email address. This, coupled with reddit's strong stance on limiting the personal information that redditors share with one another, creates a space where individuals may be willing to disclose more information about themselves with the sense that it is not tied to their "real" identity outside of reddit. Still, it may be possible to ascertain basic information about a redditor based on his/her account over time—things such as location (based on postings/comments to local subreddits), interests (based on postings/comments to other subreddits), and even intimate (though somewhat anonymized) details regarding sexual interests, relationships, and so on (which can be found by examining posts to /r/relationships, /r/sex, or anecdotes offered in threads such as /r/AskReddit). Since the user profiles on reddit also default to displaying the content a member has submitted, other members can make assumptions about one's interests or at least the subreddits which someone regularly visits (and is presumably subscribed). If aggregated, a clever redditor could determine much about the tastes, opinions, and habits of another member—as mentioned in regard to RedditGifts. And if the account in question suggests the redditor is female, other members will often look for /r/gonewild submissions (one of reddit's not-safe-for-work [NSFW] communities featuring scantily clad or nude images of other redditors) and comment on whether or not they have submitted images to the subreddit. (I discuss the gender dynamics of reddit further in Chapter 6.)

Given the ease with which new accounts can be created and the amount of potentially personal information that can be gleaned from a redditor's primary account, a common tactic is to create a "throwaway" account when discussing particularly sensitive information. For example, threads in /r/AskReddit that discuss subversive sexual habits or just strange proclivities may directly address this in the question text by saying, "Get your throwaways ready" or the like. Other times, an OP will directly acknowledge that a particular post has been made from a throwaway account or will create a username with the word "throwaway" in it, such as "Throwaway5043," with users appending random numbers at the end because so many others have also used the term "throwaway" for their accounts. While it is possible that someone would create a throwaway account without having a primary reddit account, it is unlikely, as the term "throwaway" signals to other redditors that somehow the information disclosed is not something this person would like to have associated with his/her primary reddit identity.

At the same time, there also seems to be a reluctance to "go public" with one's primary account information. Postings crop up regularly about roommates and friends discovering a redditor's username and the resulting drama that ensued or could potentially ensue. This tends to be particularly worrying when said account has posted/commented/upvoted content on

NSFW/NSFL (not-safe-for-life) subreddits. Presumably, this has to do with the redditor's desire to manage the kind of information others may glean about him, particularly if they are friends, coworkers, or employers. Worries about the general reddit public knowing something potentially damaging or unflattering seem less primary. This supports Helen Nissenbaum's (2010) notion of "contextual privacy"—that privacy is something that is actively managed and that the context, rather than the content, often is more important when negotiating privacy boundaries online.

Authenticity on reddit is less about the notion of a one, unitary identity—a direct challenge to many of the social media platforms that are far more likely to be driven by the design perspective articulated by Facebook CEO Mark Zuckerberg, who believes that someone having two or more identities "lacks integrity" (Kirkpatrick, 2010). Unlike Zuckerberg, redditors do not equate "integrity" with identity—instead, they prize authenticity, candor, and transparency. This complex notion of authenticity also extends to "outsiders" who engage with the community by posting in the /r/IAmA subreddit. As mentioned in the introduction, AMAs are opportunities for redditors to have real-time conversation with celebrities, politicians, scientists, and musicians. AMAs are scheduled and publicized weeks in advance so that community members can be ready with questions as soon as the AMA begins. Because reddit's algorithm prizes recency and because the earliest upvotes of a thread or comment count more, redditors must be ready to participate immediately if they hope for their question to gain enough upvotes to be visible to the community and AMA participant (and thus, more likely answered). The most popular AMAs reflect both the general popularity of the individual involved, and the specific appeal the celebrity has for reddit's technology-savvy/geek audience. As of April 2014, the four most popular AMAs were President Barack Obama (US politician), Sir David Attenborough (British naturalist), Bill Gates (Microsoft founder and philanthropist), and Neil deGrasse Tyson (US astrophysicist). These individuals would likely be considered popular figures on most web communities. But the top 10 also feature several reddit-specific curiosities: at number 5 is Allena Hansen, a woman who was mauled by a black bear; number 7 is reddit celebrity /u/Unidan (known for his extensive knowledge of biology); and number 8 is a man who was born with two penises. At number 17 is an AMA by a vacuum repair technician. Other popular AMA guests are also regular reddit contributors, such as rapper Snoop Lion (formerly known as Snoop Dogg), beloved by the /r/trees (cannabis) subreddit, and former US governor and action star Arnold Schwarzenegger, who often posts to /r/Fitness.

Part of what the reddit community seems to thrive on in these exchanges is the sense that members are able to access these prominent individuals without relying on gatekeepers. This may be one reason AMA participants are quick to mention whether someone else is helping them type out their responses, for example—they wish to be seen as being as transparent about the process as possible. There is an unstated expectation, too, that the celebrities will be willing to disclose and will engage with the reddit community in a way that suggests an unfiltered lack of self-consciousness. In addition, reddit clearly prizes those who will not only tolerate but also "roll with" reddit's eccentricities and sense of humor. For example, a long-standing AMA tradition is for someone to ask the person of interest whether they would rather fight 100 duck-sized horses, or 1 horse-sized duck (see more about this meme at Brad, 2013c). AMA speakers who not only answer this question but also do so cleverly or mirror the same kind of humor that the reddit community venerates (for more about reddit humor, see Chapter 5) are considered authentic and good sports. AMA participants are also expected to answer questions in a timely manner and not dodge or not respond to highly upvoted questions. When a person who planted trees for a living posted an AMA about his/her work and then proceeded not to respond to redditors' questions in a timely manner, the community was annoyed (PlantSomeTrees, 2014). One highly upvoted comment read, "Starts AMA. Makes like a tree and leaves." Variations on this response were common, despite the fact that the OP posted that he/she would be out planting all day and would respond to questions after work. While the OP might have been better served by posting his/her AMA after work instead of before, such interactions also expose reddit's sometimes unrealistic expectations for those who participate in the community infrequently.

Celebrities who do not follow reddit's unspoken expectations of authenticity are especially subject to reddit's ridicule and ire. Woody Harrelson's disastrous 2012 AMA appearance demonstrates what happens when a celebrity is misinformed or simply does not understand the purpose of the AMA format. In the AMA, Harrelson repeatedly dodged highly upvoted questions—a huge violation of the unspoken reddit norms governing AMAs—and gave terse, one-line responses that focused on the movie he was promoting. As many redditors noted, it was as if Harrelson's PR person was actually responding for the star. And many suggested that Harrelson did not understand that reddit was not for pitching movies, but "pitching oneself." One poster addressed Harrelson directly, writing,

> Mr. Harrelson, I hope you get a chance to read this. I think you were sent here under the wrong pretenses. Someone on your staff (or with the movie Rampart) likely billed a Reddit "AMA" as just another interview. That isn't what these are for.
>
> These AMAs are a chance for your fans to interact with you directly, and usually ask some pretty random questions. It's also considered bad form to make them brief, though I can imagine your time is valuable. I just wanted to apologize for the misunderstanding.
>
> If you have the time or inclination at a later date, I would highly recommend you come back to do an AMA with the right expectations. (iamwoodyharrelson, 2012)

Both Harrelson and the movie he was promoting, *Rampart*, became the butt of jokes and memes for weeks after the incident. And it became a clear warning signal to other AMA participants, and probably their PR people, that such a dismissive attitude toward reddit would not be tolerated.[9] The idea that an AMA should not be viewed as "just another interview" indicates an unspoken cynicism about the typical Hollywood press junket, as well as an expectation that celebrities will treat the reddit experience differently. The idea that a celebrity would participate in an AMA simply to "sell" one's brand is anathema.

Of course, there might be many reasons that an AMA participant would not engage with reddit on reddit's terms—perhaps because he/she feels uncomfortable with the level of disclosure many redditors request or is not comfortable talking to the public without some sort of buffer or simply does not understand the nature of the community's humor and discursive norms. But redditors seem to see such slights as personal offenses—as if the celebrity is not properly acknowledging the influence that the community has or that he/she does not understand that the community is much more savvy than the average group of fans. I would argue that there is an unstated expectation that a celebrity will both understand the "specialness" of reddit and respond to this "specialness" with a higher level of disclosure—treating redditors less as passive fans and more as collaborators in the creation and support of the celebrity's fandom community. For as much as redditors are wary of the institutions that support a celebrity or politician's success, they are equally intoxicated by the community's power to make or break someone's reputation.

Expectations of this brand of reddit authenticity extend not only to celebrities and public figures who may frequent the site but also, at points, to individual redditors. And, when reddit perceives that someone is being untruthful or inauthentic, the mob may turn its attention inward. The following is a case in point—an April 2014 "Confession Bear" posted to /r/AdviceAnimals around the time that many US college students were graduating and thinking about the value of their education (for a comprehensive discussion of the Confession Bear meme, see Vickery, 2014). Around this time, a number of image macros appeared that discussed the relative (un)importance of college education, with sentiments expressing concern that the poster had wasted his/her past four years (LogicPlacebo, 2014), with another image macro suggesting that "time isn't wasted when you're getting wasted" (I_am_one_wth_it_all, 2014) and yet another arguing the controversial opinion that if someone felt it was a waste of time, he or she probably majored in art (amlaviol, 2014). In response, a redditor created a Confession Bear image macro which read, "I earn over 200k a year / It's entirely due to my college degree" ("On the Theme of Higher Education Haters," 2014). After the post gained a number of comments and upvotes, another redditor noted a few inconsistencies in the OP's past comments that suggested he was actually a young college student. This redditor posted a Maury Lie Detector[10] image macro which read, "You claimed to make a 200K salary / Your previous comment about being an 18 year old college student determined that was a lie" (HeLivesMost, 2014). The posting was soon upvoted and reached the front page of /r/all, and within minutes individuals were downvoting every post and comment the individual made. Within a few hours, the OP deleted his account and posters in the thread were (virtually) toasting their success. Few redditors in the thread suggested that such mob mentality could have negative consequences for the OP; most argued that he should expect such a response when lying on the internet.[11] One contrarian responded on behalf of the OP, writing, "You are right. The user likely lied for karma. I still think causing hundreds of people to downvote his content is malicious and cowardly. I'm glad it's just a reddit account. I'm sorry, but I think getting karma for what you are doing is much worse." Later this individual added,

> A bunch of redditors who have likely done worse are sleeping with content smiles on their faces. These downvote brigades bother me more than they should. Maybe it's the mob mentality, but there are people behind the usernames. He might be a bad guy, but we can't possibly know that based on this one thing. Again, at least it's just an internet account and nothing too serious. (KittyFooties, 2014)

Another poster justified the downvoting brigade, arguing, "Well, if he cared so much about Reddit, maybe he should have thought just a little bit harder about lying for karma in such an obvious fashion" (EMCoupling, 2014). But just two months earlier (February 2014), moderators of /r/AdviceAnimals restated that reddit-wide rules also governed behavior in the subreddit, which includes, as one of the moderators in the comment section noted, not allowing witch hunts (colocito, 2014).

As I watched this unfold, I was struck by how quickly the reddit community turned on this young college student. If he created this macro as a kind of wish fulfillment or simply desired the approval of his reddit peers and posted a Confession Bear he knew would garner attention/karma points, what real harm was done? This was not a "Stormfront Puffin"[12] that was expressing blatantly racist views, nor did the user's account give any indication that he was anything more than a typical college student (i.e., similar to the rest of reddit's audience). Yes, he might have been simply posting something for the karma, but then so does most of reddit. So why did this person draw the community's ire? After spending three-plus years plumbing the depths of reddit, I can say with confidence: I have no idea. Or, more accurately, the only explanation I have is that reddit's cynical nature abhors a vacuum, and so perhaps individual self-loathing that many redditors seem to feel turned inward. Perhaps it was precisely *because* this person was so much like the rest of reddit—that they could see themselves reflected in the actions of this college student who was hoping to land a $200,000-a-year job after college—that he became an easy scapegoat for the reddit mob and its desire for "authenticity" in all its forms. This kind of expectation of authenticity seems to go beyond the familiar trope routinely heard on reddit (borrowed from 4chan), "OP never delivers," or its more crass expression, "OP is a faggot."

OP is a faggot

Redditors can become extremely cynical toward the world outside reddit and, often, toward their own community. This manifests in multiple ways. An ongoing theme on reddit is the tendency for individuals to ask for help with a difficult problem or offer a tantalizing story that hasn't come to a conclusion yet, sparking speculation and interest from the community, but then never present a resolution. If these posts get upvoted to the front page, they will often provoke long chains of comments asking for clarification from the original poster (OP) or relaying similar anecdotes from other posters. At some point, these conversations tend to turn into a discussion of whether or not the OP will actually follow up on his/her promise to "deliver," that is, finish the anecdote or story, with meme responses often offering hyperbolic scenarios whereby reddit observers will die while waiting for the OP's conclusion. Given the amount of time that many posters invest in commenting on these sorts of threads and indeed the material and emotional resources they may leverage on the OP's behalf, there seems to be a reasonable expectation that the community "deserves" some sort of return on members' investment, in the form of a new post updating the situation or at least an edit to the original post with news.

A concrete example helps illustrate the numerous ways in which these moments demonstrate the fascination reddit has with past events on the site. In early 2013, one redditor posted a picture of a large safe he/she had found in a newly purchased home, which resulted in a number of comments, and a long, detailed discussion of the various ways the poster might go about cracking it (dont_stop_smee, 2013). Speculation about what the safe contained abounded and resulted in humorous threats that the individual would not offer a follow-up.[13] And after some time passed, it was clear that the OP did not intend to follow up. The idea that OPs should "deliver" suggests an intense desire for reciprocity within the reddit community. As a result, certain subreddits have rules (stated or implicit) that encourage redditors to provide updates to their situations. The subreddit /r/relationships, dedicated to requesting and receiving relationship advice, encourages individuals to update their situation in a new post after 48 hours have passed. However, the entire safe incident spawned a new subreddit, /r/WhatsInThisThing, which offered a dedicated space for individuals to post images of some locked object and invite others to speculate on how to open it and what the object might contain.

GIF and text-based responses to these kinds of situations regularly include the 4chan-borrowed meme, "OP is a faggot" (BraveSirRobin, 2011). This offensive slur is supposed to be read as a "humorous" remark about the tendencies for OPs not to follow through with their promises to update the community. It is likely intended more as a form of policing behavior and less an actual statement about sexuality (Pascoe, 2011). Using homophobic language as a linguistic shortcut seems both acceptable to a large number of redditors (as evidenced by the fact that these comments, while not usually the most popular in the thread, have a substantial number of upvotes and related comments) but also unsavory or problematic by others, as the use of the "OP is a faggot" phrase appears regularly in threads about things that redditors hate most about reddit. It also contradicts the image of the prototypical redditor as socially liberal—or perhaps it reflects the same ethos that leads to the "hipster racism" (Kerckhove, 2007), which often permeates redditors' use of the term "nigga" or its even more offensive counterpart.[14] Likewise, the words "retard" and "retarded" are also bandied about to describe someone who is acting weird or to point out that something is nonsensical. Redditors' generous use of all three terms seems to suggest that many of them believe that racism, homophobia, and ableism are concerns of earlier generations and/or not relevant to their day-to-day experiences (Haywood, 2012; s.e. smith, 2009a, 2009b). (In the meta-subreddits, such comments are often referred to ironically as "edgy.") And perhaps because they recognize the implicit homophobia in the use of the term "faggot," redditors also employ a different strategy that still recalls the word by referring to OPs as "a bundle of sticks." Still, the frequent use of these kinds of slurs—even while decried by a portion of the reddit community—reifies the image of the redditor as privileged and relatively unaware of the power of the language they use.

How redditors view themselves

As a community, reddit has an unsurprisingly mixed relationship with itself, especially as the site has gained popularity and become more visible to the outside world. However, the cynicism that characterizes at least a part of reddit's culture is not directed only at institutions or those outside of reddit. It is also directed inwardly—at both the community as a whole and individuals therein. Sometimes it takes the form of humor, as in the case of the Socially Awkward Penguin image macro.[15] Or the "Scumbag Reddit" image macro, which expresses dissatisfaction with some aspect of the reddit "hivemind"—or the tendency for subreddits to become echo chambers (a point I discuss further in the next chapter). Many of the site's meta-subreddits, such as /r/SubredditDrama, which engage in meta-commentary about the site's interactions and the kind of individuals it seems to attract, often lapse into discussions about reddit's curious behaviors as if they are some kind of alien tribal culture.

When thinking about how redditors see themselves, models from popular culture are often invoked. One of the quintessential prototypes of a redditor is Sheldon from the US television series *The Big Bang Theory*. Sheldon is an awkward genius on the show who often fails to understand the basics of human interaction while still being a scientific phenom. Even the show creators imagined Sheldon as being the kind of person who might frequent reddit, as he was shown wearing a shirt with reddit's logo (the alien "snoo") during one of the show's episodes. Redditors noted this fact and debated whether or not Sheldon fit the profile of the "typical" redditor. Many of them argued he would not be a "redditor," as well as noted their general antipathy toward the series. As one poster astutely argued, however, Sheldon exemplified the essence of reddit:

> It's funny how people can believe that Sheldon wouldn't be a redditor. A place where a majority of condescending, awkward, know-it-alls, circle jerk over each other about how they are intellectually superior to those who are not as informed on subjects at hand or do not share their own moral perspectives. While also having endearing qualities like fondness of cats (I'm not just talking about the soft kitty song but there was an episode where he owned several dozen cats).
>
> Sheldon is not only a redditor. Sheldon IS reddit. It's only too ironic because Sheldon would probably hate the big bang theory too because it doesn't portray "nerds" correctly. ("Big Bang Theory Irony," 2012)

This poster notes the inherent, often unspoken self-loathing that redditors have about themselves and the community at large. Even wearing a reddit T-shirt or identifying as a "redditor" is viewed with some amount of disdain. A popular posting to AskReddit asked individuals to identify something that "screamed douchebag" when worn. One redditor responded, "A redditor t-shirt. It lets me know they are probably a judgmental douchebag …" to which others agreed, and added, "Who would honestly wear a reddit shirt in a public setting?" (bennyschup, 2014). Another contributor in the same thread went on a long diatribe condemning anyone who would deign to wear a reddit T-shirt:

> It tells me that the person is probably a fucking loser, most likely a college drop out who thinks they're a know it all and probably someone who just parrots everything Jon Stewart, Bill Maher and MSNBC tell them to think. They're most likely overweight, insecure and single, lacking in almost every single way when compared to their contemporaries …
>
> If the typical Redditor was an animal, it would probably be a rabbit without back legs. Harmless, useless and clinging to life, just barely. I'd want to put it down, do it the favor of ending its pathetic existence. That's what I see when I think of Redditors in the real world and people who would actually wear Reddit clothing. (KidsAre9532, 2014)

Others in the exchange tried to figure out whether this person intended these hyperbolic statements to be read ironically, while others wondered why he still frequented reddit if he had such a negative view of the community. Another individual argued there was a fundamental difference between those who used reddit instrumentally, as a link-sharing/news-sharing site, versus those who "try and make an identity and community out of it." These kinds of exchanges are not infrequent, suggesting that there is an inherent tension between those who imagine reddit as a community and those who view it as merely a platform. Still, I would argue that this kind of (1) distinction and (2) meta-talk about the nature of reddit is unique to this space, especially as compared to other participatory culture platforms and social networking sites.

In a more recent posting on /r/videos, a contributor suggested *30 Rock*'s Liz Lemon could be considered a typical redditor: awkward, sarcastic, and brainy (Sergnb, 2014). Others argued that *The Office*'s (US) Dwight Schrute was a more accurate depiction, with his outrageous conspiratorial ravings, adherence to rules no matter how inane or pedantic, and constant attempts to demonstrate his superiority. Still other posters took issue with the entire thread, suggesting that reddit was far too quick to suggest that it is only *other* redditors who are nerdy, socially awkward, or self-important. As one individual wrote, "Man this thread is filled with pedantic assholes. 'Everyone on reddit is the comic book guy from the simpsons except me'" (Sergnb, 2014). Another poster responded to suggest that this was precisely the essence of reddit:

> Well, yeah, but that's basically reddit in a nutshell, isn't it?
>
> Nobody hates redditors and redditor behavior more than redditors, and approximately 95% will gladly tell you about how they're the exception to the rule. (The other 5% just don't feel the need to tell you, thinking it self-evident—since they're making posts talking about how all these other redditors will think these things.)
>
> Everybody claims that they don't give a shit about Karma, and everybody acts in a manner indistinguishable from people who really, really care about Karma.
>
> Everybody hates circlejerks, except when it's one they're a part of, in which case, it's not a circlejerk, it's just a conversation between like-minded people.
>
> Everybody has a story about how they bravely went against the hivemind and got crushed with downvotes but still stayed true to their statements, and about that fucking idiot who was just being contrary and deserved every downvote they received.
>
> Reddit is a weird, weird place. Or maybe it's utterly normal, who am I to tell? (Churba, 2014)

This kind of intense scrutiny of the "reddit community" is a regular occurrence on the site—with redditors regularly decrying the groupthink/hivemind that stifles alternative perspectives. Perhaps it is the fact that the site has become the hub for a particular kind of geek masculinity (a point to which I return in Chapter 6) that creates a space in which individuals regularly assert their "outsider" status. Redditors often view themselves as outside the mainstream or being counterpoised to traditional hegemonic notions of masculinity, and yet the algorithmic nature of reddit's upvoting/downvoting system means that it can easily seem very "insider-y," with alternative or nonconforming opinions often swimming in negative karma.

I asked my informants who they thought was a "typical" redditor. Many of them were hesitant to answer, suggesting that they did not want reddit to be characterized as a monolith or having only a single kind of person as a member. After providing this caveat, however, most of my informants landed on a strikingly similar composite image: a geeky, male atheist who is college educated, from the US, and interested in STEM-related disciplines. One informant, Marley, wrote, "The typical redditor is a young adult male, middle classed, with some education. Perhaps a little on the nerdy or geeky side. Probably from the USA, Canada, or Western Europe." Amari agreed but was reluctant to describe a typical redditor for fear of lapsing into stereotypes: "If forced to stereotype, I would say an American, European or Australian millennial male with a couple more years, 18–26, either in college, just out, or working. I think there are pretty vast groups of older (than average—maybe 35–50) people, women, minors, and people with VERY low educational attainment. The prevalence or horrible and overt racism and sexism indicated that to me." Dakota echoed the demographic profile listed above but highlighted some of the other, non-demographic connections that bind the community together:

> I think the typical redditor is well-informed about his own area of interest. This can come across as a) actually knowledgeable (in his area), b) cocky but wrong (not in his area), or c) creepily obsessed with his area, depending on how he expresses himself. Redditors spend "too much" time on the internet because the internet is a community for them—which is an interesting, legitimate social shift in the 21st century with globalization and access to technology. Redditors also tend to be bored, or at least unenthralled by the real world, leading them to resort to expanding their horizons or finding new interests online. I think that it's hard to use reddit (and other social media forums and news sites) without becoming a bit addicted or over-involved.

Jamie noted diversity as being the hallmark of the "typical" redditor:

> The great thing is, there is no "typical" redditor. People come from all walks of life, from all kinds of jobs and careers, from all kinds of educations and upbringings. There

are men, women, lesbians, trans, lawyers, dentists, students, people who ride the rails. There is no such thing as Typical on Reddit.

This is true—reddit does attract a variety of people, especially to its smaller, niche subreddits. However, the large majority of individuals who frequent reddit are likely to reflect both demographic and attitudinal realities. The most comprehensive answer I received when asking individuals about this is reproduced below in its entirety. It mirrors my own image of the "prototypical" redditor who likely frequents the default and more mainstream subreddits after three years of fieldwork in this space. River wrote,

> I'm finding it difficult to write what I think about the demographics, beliefs and attitudes of the "typical" redditor because I am self-conscious about how negative it might sound. I also don't want to over-generalise. Reddit has a huge userbase, and there's no way of telling (for example) what the demographics of readers or more casual users of reddit are compared to those who might be engaged enough to comment in various polls. Different subreddits also have different demographics, but, since you want my opinion, I should bite the bullet, so, here goes.
>
> The typical redditor is white, male, 22 years old, from a middle-class background, a natural-born citizen of the USA and studying a STEM subject at university. He is interested in internet culture, video games and hollywood films. He reads, but not excessively, and mostly on the internet rather than books. He voted for Barack Obama, but is probably critical of his administration. If he is engaged politically, it is most likely to be in the form of signing petitions or participating in other sorts of "armchair" activism. He may donate a small amount to some cause or be involved with raising money on campus, but not to a great degree. He is more likely to be concerned with causes that have some sort of scientific bent, such as curing cancer or promoting space exploration.
>
> He is an atheist, quite possibly a "new atheist", is deeply critical of religion, and has a powerful belief in the concept of progress, and that the route to progress is scientific and technological advancement. He sees all religion and whatever else he considers to be "unreasonable" as road blocks towards this, which colours his political and social thinking. He is likely to (for example) prefer criminological rather than sociological explanations for crime and hold the conclusions of evolutionary psychology with great respect. His tendency to look at things like race, gender, intelligence and so on as rooted in biological facts puts him distinctly at odds with schools of thinking such as feminism, marxism, post-modernism and their variants, which he considers to be all nonsense. He prizes education and knowledge highly, so long as it falls within his range of interests and beliefs, and he probably considers studying humanities at university level a waste of time and money, in terms of getting a job if not absolutely. His atheist beliefs also shape his attitude towards geopolitics, and contribute to an attitude which, whilst recognising the "war on terror" to be in some way a mistake, preserves a healthy fear of the potential dangers of Islamic fundamentalism.

> His attitude towards relationships is shaped by his belief that, in many ways, men and women are fundamentally different, and thus have different needs and desires. He does not consider himself a misogynist, but he is likely to prefer the company of men and views stereotypically feminine things with an outwardly good-natured disdain. He uses pornography and would consider hiring the services of a prostitute, and views neither actions as things that should ethically concern him personally.

What strikes me about River's response is how thoughtful, comprehensive, and self-aware it is. The irony, of course, is that the individual who produced this spot-on description of at least the image one might have of a typical redditor given the attitudes frequently expressed on the site is also a redditor! As am I. As are the denizens of many subreddits who likely do not fit this image at all. Reddit is, as some might suggest, merely a platform, a set of tools, and an algorithmic way of sorting information and interacting with others around said information. Still, I believe it is extremely telling that my informants, discussions of reddit and redditors on some of the meta-subreddits and in the defaults, and my own observations all coalesce around the same some sort of composite image of "a redditor." So dismissing it as *merely* a platform is not entirely accurate, because platforms themselves have politics, as do the communities that engage with those platforms. And even the self-loathing that redditors may express about others who also contribute to these spaces is telling and might possibly reflect their inherent cynicism and skepticism about institutional structures more broadly.

I think it is important to note, too, that these images of the "redditor" coalesce not only in whom they include but also, notably, in whom they exclude. In both my fieldwork on the site and in discussions with my informants, I have been struck by the lack of diversity in the kind of redditors described. Redditors are not envisioned, in the broadest terms, as female, older, humanities-minded, activist-oriented, working-class, people of color from a developing country, nor are they (generally) disabled or late adopters of technology and/or popular culture.

Rewarding the authentic; punishing the disingenuous

The multiple ways in which redditors negotiate their relationship with each other and with themselves suggests a complex desire for authenticity and specific, individual connection with others. Through RedditGifts, Reddit Gold, and other small and large acts of kindness, redditors express their affection and appreciation for others and reward authenticity. At the same time, they harbor a distrust of large institutions, which at moments can even include the reddit community itself, and are especially wary of those who either don't recognize the community's potential power or are unwilling to engage with reddit on reddit's terms. One of the key hallmarks of reddit is the implied desire for candor as a mark of authenticity, or at least the appearance of candor (Madrigal, 2014), which I argue extends both to reddit members as well as those who engage with the community through AMAs. Public figures who are unaware of these expectations or refuse to accept them as conditions of their participation are often mocked, derided, and/or punished for their transgressions and ultimately perceived as disingenuous. And then there are those private individuals caught in the crossfire—unintended casualties of reddit's mob mentality that can be both amoral and vicious. And, while reddit can bring together a vast, diverse audience, many redditors envision a "prototypical" redditor who often represents many stereotypes of geek culture. He (and it is invariably a "he") is socially awkward, is romantically challenged, and demonstrates intense interest in all things sci-fi, gaming oriented, or STEM related. While this image serves as a humorous trope for subreddits such as /r/circlejerk, it likely also disenfranchises would-be redditors who are less likely to "see themselves" on the site's /r/all.

Notes

1. http://redditgifts.com/gallery/
2. Individuals can also send a postcard to reddit's headquarters in exchange for a month of Reddit Gold.
3. There are numerous kinds of cryptocurrency in use on reddit. Bitcoin and Dogecoin are the most popular, but there are around 36 other cryptocurrencies on reddit, including Anoncoin, Einsteinium, Flappycoin, Litecoin, Nyancoin, ReddCoin and FedoraCoin (see the sidebar of /r/CryptoCurrency for a more comprehensive list).
4. https://www.youtube.com/watch?v=GOlTM297CFE
5. http://www.youtube.com/watch?v=rDZf4A9X3W0
6. http://www.youtube.com/watch?v=zUQZsIJPDrw
7. http://mankabros.com/blogs/god/category/tween-jesus-me/
8. Of course, this is merely my own speculation and is worthy of a comprehensive study to determine its validity.
9. See, for example, Bronson Pinchot's tongue-in-cheek introduction in which he mentions "no questions about *Rampart*" (BronPinchot, 2012), Jared Leto's postscript to his AMA introduction 18 months after the Harrelson debacle, "don't forget—RAMPART! on VOD [video on demand] now!" (_JaredLeto, 2013), and reddit-favorite Zach Braff's 2014 AMA

titled, "I AM Zach Braff. Ask Me Anything. (About Rampart)" (zachinoz, 2014). Reddit does not forget.
10. This image macro features a screen capture of US talk show host Maury Povich as he reads from a lie detector test (a recurring occurrence on his show) (Don, 2013a).
11. I find this *highly* ironic given that a popular comic lampoons this perspective and is regularly posted to reddit in threads such as these. It borrows lines from the PBS children's show *Arthur* in which one of the characters says, "You really think someone would do that? Just go on the Internet and tell lies?" (Brad, 2013d).
12. "Stormfront Puffin" refers to the "Unpopular Opinion Puffin" image macro that expresses a supposed opinion that is unpopular or (more often) actually quite broadly accepted by the reddit community, but perhaps "unpopular" to others. "Stormfront Puffin" is the term used by some segments of the reddit community who have noted that the Puffin often expresses racialized stereotypes and is (according to some within the social justice–oriented subreddit /r/ShitRedditSays) often a front for white nationalists from the Stormfront organization who are using reddit as a recruiting ground. And, unfortunately, these macros were often upvoted to the front of /r/AdviceAnimals by unknowing redditors—until they were banned in May 2014. See more about reddit's problematic discourse in Chapter 6.
13. An earlier safe-opening incident led to the OP never following up with the reddit community, despite asking for (and receiving) a significant amount of help (secretsafe, 2011). Thus, this particular thread was filled with statements like "oh, here we go again …" to indicate the unlikelihood that the OP would update its contents.
14. Redditors' proclivity toward hipster racism, homophobia, and ableism are by no means encapsulated solely within the use of these terms—instead, I offer them here only as the most obvious forms of this kind of ethos that permeates many interactions on the site.
15. The Socially Awkward Penguin (SAP) image macro revolves around awkward moments that the author has been a part of or has witnessed (James, 2014). For example, the meme might say something like, "Server says, 'Have a nice meal.' 'You too.'"

· 4 ·

RELIVING THE PAST [REPOST]

Redditors are a curious bunch. For all of their excitement about discovering, sharing, and reveling in the new, they also spend a lot of time talking about the past. Stumble into any public conversation on the default subreddits and inevitably someone will make some reference to a past meme or a past event that is from reddit's history or suggest the post is actually a repost or decry how crappy the content has become since he joined the site. He retreads the same conversations, posts and remixes the same memes, and then complains about both. This becomes part of the (in)famous "hivemind" or "reddit circlejerk" of which many redditors disparagingly talk. Given all of this, why would anyone keep visiting, voting, and commenting on reddit? For many, the answer is that they unsubscribe to the defaults and ignore whatever is happening there. These redditors stick to a smaller subset of subreddits tailored to their interests that mostly avoid these kinds of continual replays of the same conversations. Part of this transition, I argue, happens as site visitors move from being novices to becoming full members of the community.

Others take a different path. Some individuals demonstrate an almost myopic focus on reddit's own inner workings, its drama, and its eccentricities. As I mentioned in the last chapter, redditors have a preoccupation with themselves that can manifest in both cynicism and a kind of naïve altruism, and the community often cycles between these two dialectics at astonishing speed. The "reddit mob" is not simply a just one—mobs rarely are. But it is not wholly lacking self-awareness; like most things redditors do, it is subject to great self-scrutiny and even moments of empathy. And this is one critical aspect that separates the community that gathers on reddit from other participatory culture platforms—it engages in a tremendous amount of meta-talk about the nature of the community itself.

Community membership—the transition from newbie to "redditor"

Scholars within computer-supported cooperative work (CSCW), education, and organizational communication have long been interested in the ways in which individuals move from novice to experts within communities. Etienne Wenger's (1998) work on communities of practice is particularly influential in this regard. Wenger grounds his discussions using social learning theory as a foundation for understanding the ways in which novices learn how to be experts—it is, he argues, a fundamentally social process that happens in relation to others within the same community of practice. In his work with Jean Lave (Lave & Wenger, 1991), Wenger articulates the core mechanism by which this process occurs, which he terms "legitimate peripheral participation" (LPP). To become accepted members of a community, individuals participate peripherally, that is, in situations that approximate full practice but where the penalties for failure are mitigated or lessened. Lave and Wenger (1991) emphasize that the community must allow new members to be considered for "full" participation—meaning the space cannot be so closed that there is no way for novices to learn from the inevitable mistakes they will make. The classic example of LPP is the apprenticeship model, in which a novice spends much of her time observing the practice of a master craftsperson and is given low-level tasks to complete (like sweeping in a woodshop), gradually gaining additional responsibility for particular aspects of the process (such as kiln maintenance and firing in a pottery studio) until she becomes a full-fledged member of the community.

While the period of acculturation to reddit is much less intensive than it would be for a law student learning the ins and outs of legal jurisprudence by clerking for a district judge, for example, there is a process by which individuals become contributing members of the reddit community. Typically the process goes much like my own experience described in the introduction, with redditors often lurking for a time before contributing. Assuming they actually choose to create an account (which, given the site's numerous page views versus the number of accounts reddit boasts, is actually rare), a number of redditors I talked to often lurked and voted on content in their subreddits of interest for some time before actually submitting anything. It is relatively common to see comments by a new redditor suggesting he "delurked" and created an account just to contribute to that particular discussion. Whether or not this is actually the case or just a clever ploy for karma is hard to say. Still, despite this low technological barrier to account creation, there seems to be a sense that many redditors lurk for some period of time, perhaps only voting, before submitting actual content. Other talk on the site suggests that sometimes redditors bluster in by posting something to a subreddit without understanding its norms and are downvoted or just plain ignored as a result. Unlike the typical apprenticeship model mentioned above, reddit has no formalized process by which novices learn how to be "redditors" from more senior community members. However, interactions with moderators might serve a similar purpose, if, for example, someone posts an inappropriate comment or submits a link to the wrong subreddit or breaks a subreddit's rules. Votes also serve to "teach" a novice how to interact with the reddit community—presumably, negative karma offers would-be redditors a sense of what is and is not acceptable in the subreddit with which they are interacting.

When discussing how they came to find out about reddit, I was surprised how many of my informants were introduced to the site by a friend or significant other. Other informants came to the site after hearing about it mentioned in the media (a new influx of users joined reddit after President Obama's AMA, for example) or after they saw content from reddit linked to from other web sites or social networks such as 9Gag or Facebook. Many of them mention feeling intimidated upon first joining, as they were not sure how exactly to interact with the community. Like my informants, given the insider-y nature of some of the conversations that go on and the quick wit that some redditors demonstrate, I also hesitated before submitting content rather than simply browsing and voting. It took a significant amount of courage for me to submit my first posts and comments, and I admit that I refreshed the page numerous times the first few times (okay, all the time) I posted to track my upvotes and downvotes. Over time, I grew both more comfortable and much more cynical about reddit—yes, all of the humor, playfulness, and intelligent discourse was there, but I was also attuned to how much of it was buried beneath what seemed like an avalanche of pointless memes and phrases, circlejerk copypasta, and problematic discourse around issues of gender, race, and sexuality. And this was especially true when I spent time on reddit's largest subreddits—the dreaded defaults.

Why everyone hates the defaults

When a casual browser of reddit becomes a true "redditor," it is likely that his perspective on the default subreddits changes. The defaults are the target of considerable scorn and meta-talk and viewed with a fair amount of disdain by many, usually longer-term, redditors. My informants noted that one of the driving forces for them to create an account was not only to be able to customize their reddit experience but also to remove the default subreddits from their front page. The defaults remain, however, the most popular subreddits, the most highly trafficked, and the most public face of reddit—as content posted on the defaults is far more likely to receive enough upvotes to make it to the front page of /r/all. Yet they are also viewed as appealing to redditors' basest desires, featuring the lowest common denominator in terms of humor, content, and interaction. And the general rule of thumb is that as soon as a subreddit becomes a default, it rapidly descends into a cesspool of uninteresting content, poorly moderated conversations, and egregious examples of the reddit circlejerk. But why is this? I have three different explanations and examples of how defaulting can change a subreddit's tenor.

The first and most obvious explanation is that the sheer increase in the number of individuals submitting and voting on content decreases the quality of material in a given subreddit. This is the case with the recently defaulted /r/food. A catch-all community for any number of culinary interests, prior to its defaulting /r/food featured a smattering of photographs and associated recipes of food that redditors had recently made, with discussions ranging from basic questions about ingredients or praise for the OP's ingenuity or discussions about culinary technique. After defaulting in May 2014, the front page of /r/food morphed into something out of a Guy Fieri menu (Wells, 2012), with numerous postings of bacon-wrapped jalapeño bites, McGangBangs (a McDonald's secret menu monstrosity made by shoving a McChicken sandwich inside a double cheeseburger), and macaroni and cheese featuring eight different kinds of cheese. Discussions on the subreddit likewise declined, with some longtime subscribers decrying the uptick in over-sauced and over-cheesed bacon atrocities, but many more, newer contributors were happy to extol the virtues of gluttonous meals that displayed little culinary inventiveness. A similar pattern of decline occurred in /r/philosophy in both the kind of content submitted and upvoted after defaulting. While not the most serious-minded philosophy community on reddit (that would likely be /r/AcademicPhilosophy), the day after defaulting, /r/philosophy featured an article about scientist Neil deGrasse Tyson (reddit's version of a patron saint), which was both highly upvoted and had relatively little to do with the field of philosophy. Both declines, I would suggest, were prompted by a rapid increase in the number of subscribers, along with a concomitant increase in visibility.

The second explanation is implicated by the first. With a rapid increase in subscribers, defaults must quickly step up their moderation of the community and "hire" more moderators to keep up with the inevitable trolls, spam, and all-around crappy content that redditors will submit. Defaults often have limited moderation policies—for example, /r/videos does not allow "political" content but has no stated policy regarding sexist, homophobic, or racist comments and postings. Thus, /r/videos often attracts the worst of redditors' nascent or outright active racist and sexist behavior. Subreddits that might have had limited problems with spam or required few rules to keep material on topic in the past (and thus, few moderators) are often underprepared to deal with the onslaught of new users who have little experience with the community's unstated norms. Inevitably, this both waters down the content and also places a heavier burden on a subreddit's moderators. Since many moderators moderate multiple subreddits, defaulting of one might mean a significant uptick in moderator burnout, resulting in even more work for the moderators who stay on through the transition.

The third explanation is that subreddits gain a significant amount of visibility if they are made defaults. Moderators also gain a significant amount of power and notoriety, which is why the leadership of many smaller subreddits might be willing to have their community default despite the increased workload. During the reshuffling of the site in May 2014, a subreddit dedicated to women's issues, /r/TwoXChromosomes (2X), defaulted. Suddenly, this heretofore relatively unknown subreddit was thrust into the reddit home page spotlight—with disastrous results. Reddit is not known as the friendliest place for women (for more discussion, see Chapter 6), and the defaulting of 2X was clearly an attempt by site administrators to give /r/all some semblance of gender balance and demonstrate to new users that there was a place for women on the site. However, the defaulting of 2X was disastrous.

Longtime subscribers were unhappy that their formerly low-key subreddit was in the spotlight and attracting much more attention from both trolls and anti-feminist men's rights folks, who were particularly angry that /r/MensRights had not been considered for default status. Moderators were accused of agreeing to make the subreddit a default just because of the notoriety it would mean for them. They responded with relatively little compassion when 2Xers started complaining about the increased numbers of harassing private messages they were receiving, instead stating that things would "calm down" as reddit grew used to having a subreddit for women's issues on the front page.[1] In this case, the increased visibility that default status brought 2X made it a target for many different groups that were not happy about its heightened profile.

Is hatred of the defaults warranted? Clearly, if they did not appeal to some segment of the reddit audience, they would have far fewer subscribers than they do—even assuming every new account is subscribed to them by default. The defaults do serve an important purpose from the administrators' perspective, as they introduce newbie users to a small segment of the vast diversity of content one can find on reddit. This is why the radical change in 2014 from featuring 25 defaults to 50 was such a big deal—it demonstrated a desire by administrators to change the home page from one dominated by postings from "humorous" image-heavy content, as well as technology- and science-oriented material, to one that might appeal to a broader audience. Hence the inclusion of subreddits like /r/writingprompts (where redditors post potential story ideas), /r/mildlyinteresting (a lower-key and much more interesting version of /r/funny or /r/pics), /r/creepy, and the aforementioned /r/food and /r/TwoXChromosomes.

As mentioned above, many of my informants noted that being able to unsubscribe from the defaults became one of the major motivating factors for creating an account on the site. As Marley noted:

> One thing that I couldn't do with out is the ability to filter subreddit subscriptions to fit my personal interests and tastes. I really don't understand how people can "lurk" reddit without creating an account. There is so much "noise"—mostly generated by bored young people—in the default subs that the content I come to reddit for is usually completely buried. I really cringe when I see the front page before I log in to my account. This brings up my previous point about reddit not being for everyone. As it stands, you really need to put time into tweaking your submissions to get a pleasant and informative reddit experience.

Also, some informants, especially those who serve as moderators for one or more subreddits, argued that moderation is extremely important for keeping conversations on all subreddits on track. But it is a tricky job, as the 2X case demonstrates, as moderating with too heavy a hand can alienate much of your audience. And it is made especially difficult when the tools moderators have at their disposal are insufficient for doing a good job. As Harper offers:

> It's really really really difficult for a mod to try to change their own userbase, especially with Reddit's minimal moderation tools. Having clear and well-enforced rules stops the crap from reaching the front page, but actually trying to cut down on the amount that gets submitted is tricky.
>
> Subreddits that are under-moderated will have issues with a high volume of low-quality submissions because when subscribers are not aware of standards they submit more shit. Subreddits that are over-moderated risk alienating their userbases or being slow.
>
> Moderating for quality is really, really tricky anyway. Most users would shit their pants at the idea of a mod removing a post just because it sucks. Still, without a high level of attentiveness from the modteam and a resolve to keep quality at a certain level, quality will be horribly low. (think /r/atheism pre-shakeup). Some subs are too big to be saved, like r/funny and r/pics, but at the very least mods can prevent subs from being worse than they could be.

In a subreddit with over 6 million subscribers (as /r/funny has), it is nearly impossible to keep the quality of material high. This is both because default subreddit subject matter is often overly broad and because the defaults are the most likely initial starting point for novice redditors, who may submit a lot of inappropriate content or reposts just because they are not familiar with the site's norms. Again, as in the case of /r/funny, the subreddit rules provide little guidance concerning what constitutes "funny" material, just a statement in its sidebar: "you must be funny to submit."

Reddiquette and its discontents

Reddiquette is the main, informal articulation of the ways in which redditors engage with one another in this space. Novices are encouraged to read and internalize the "rules" of reddiquette when they join the site, and more experienced redditors are also periodically encouraged by site moderators and administrators to give it a reread. Ironically, or perhaps totally predictably, reddiquette is mostly ignored by many community members while still being a core tenet of many subreddits' rules and a point of discussion, especially in regard to how individuals vote on site content.

Unlike other participatory culture platforms such as Twitter, YouTube, and Tumblr, which provide few rules around user behavior other than those articulated in their Terms of Service, reddiquette is a lengthy set of guidelines authored by the community.[2] I believe this unique approach—both the creation of overarching guidelines for community discourse as well as the ability for reddiquette to be publicly edited—encourages an ongoing discussion of reddit by redditors. It also further shifts responsibility for "policing" objectionable, inappropriate, or even simply off-topic content onto the backs of community members and moderators. By distancing themselves from the content produced and shared on the reddit platform, site administrators not only encourage a feeling of collective ownership of the reddit space but also limit their liability and responsibility regarding offensive material.

Reddiquette covers a wide variety of interactions on the site. Broadly speaking, elements of reddiquette can be broken down into guidelines covering three categories: submitting, voting, and commenting/interacting with others. In terms of submitting content to reddit, redditors are encouraged to search for duplicates before submitting to prevent redundant postings, link to persistent URLs instead of using TinyURL or other shorteners (presumably to ensure the "permanence" of the linked materials), use NSFW tags where appropriate, make notes of any edits they make after the submission is posted, and submit a broad diversity of content (not just material from one's own blog or site). Reddiquette also discourages redditors from editorializing their submissions, asking for upvotes, reposting deleted information, or posting to subreddits that do not fit the submission (for example, posting a photo of a cat to /r/videos). Reddiquette encourages redditors to vote thoughtfully—the guiding rule being to upvote if something contributes to the conversation, and to downvote if it does not. Most important, redditors are cautioned not to simply downvote material they disagree with; again, reddiquette states,

> [Please don't] downvote an otherwise acceptable post because you don't personally like it. Think before you downvote and take a moment to ensure you're downvoting someone because they are not contributing to the community dialogue or discussion. If you simply take a moment to stop, think and examine your reasons for downvoting, rather than doing so out of an emotional reaction, you will ensure that your downvotes are given for good reasons. ("reddiquette—reddit.com," 2013)

Redditors are also encouraged not to mass downvote/upvote individuals based simply on who they are (a point I return to briefly in the next chapter) and not to report postings unless they are breaking a subreddit's stated rules; they are encouraged to read/view the material thoroughly before voting on it. Reddiquette, like the "typical redditor" I discussed in the last chapter, valorizes rationality over emotionality.

My fieldwork observations and interviews with individual redditors suggest that (1) a surprising number of them are not familiar with reddiquette or have heard the term but know very little about what it actually entails; (2) many disregard reddiquette when it comes to voting; and/or (3) they personally espouse to follow reddiquette but think that most others do not. One informant, Dylan, argued that reddiquette was "completely ineffective and ignored in most places" but added, "some smaller subreddits enforce it more actively or take greater steps to create a community where it's self-enforced." Sage agreed, suggesting that most folks disregarded reddiquette when it came to voting, but added that a few of reddiquette's main points were generally followed by the community as a whole. Specifically, this individual noted that personal information is rarely posted and argued that the quality of conversation was better on reddit, as "one word/utterly useless responses (lol/this/first/ttt/etc) are much less common than on any other online forum I've ever frequented." Still, my informants agreed generally that the *idea* behind reddiquette's voting guidelines was a good one, in that it could potentially create interesting discussions—especially ones that might challenge one's own views.

I have seen most, if not all, of the guidelines reddiquette offers disregarded numerous times during my time on reddit. Individuals are often subjected to downvoting brigades (as was the case with the false Confession Bear posting mentioned in the last chapter), in which entire submission histories will be systematically downvoted by a group of redditors. I have broken probably all of the guidelines around voting that reddiquette outlines. I have downvoted comments made by people I have tagged in RES as racist or as trolls; likewise, I am sure I have upvoted some redditors' submissions just because I have enjoyed their contributions in the past. I have most assuredly downvoted material I disagreed with despite knowing that downvotes are supposed to be assigned only when someone is not contributing to the conversation. In the abstract, how one is supposed to assign upvotes and downvotes according to reddiquette is sensible, but in practice, I found it extremely difficult to live by these rules. When coming across a comment that was well articulated but still expressed a borderline racist or "hipster racist" opinion on /r/chicago, for example, I downvoted without hesitation. When someone in /r/funny began a discussion on evolutionary psychology as a reason that women and men differ, I downvoted. And, after months of seeing misogynistic content hit the front page of /r/all in multiple forms (which I detail in Chapter 6), I started downvoting this kind of content with abandon, even if it "fit" the subreddit in question. For example, I downvoted a GIF submitted to /r/gifs featuring a tennis player presumably talking with her coach about her game that was subtitled to suggest she was agreeing to anal sex. Yes, as a submission to /r/gifs it was not breaking any subreddit rules—it was not NSFW as it featured no explicit language or images, nor did it break the other rules stated by the moderators. But, in my mind, having this kind of content on /r/all further alienates a large segment of the reddit audience. However, by the strictest standards of reddiquette, I am probably a bad redditor. I suspect I am not alone, however.

As a heuristic for evaluating whether or not to up- or downvote, reddiquette falls short. I found myself struggling with the essence of the "upvote if it contributes, downvote if not" guideline for several reasons. The nature of the reddit front page makes it difficult to know what subreddit to which something has been submitted, as the name of the subreddit appears in much smaller text than the title of the posting and is easy to overlook. This makes applying the rules of a given subreddit difficult. For example, /r/videos has no stated policy against racist or sexist content but draws the line at political videos. Conversely, /r/politics disallows "hateful speech" but specifically focuses on political content (hansjens47, 2014).[3] Identifying which subreddit one is browsing can be difficult, especially on mobile devices that limit the extensive CSS customization that moderators may use to differentiate their subreddits. Therefore, it can be easy to break a subreddit's rules unintentionally.

Reddiquette seems a holdover from a time when posts and comments could be much more easily assessed as either "contributing" or "not contributing." In the days before comments and subreddits were introduced, when reddit was strictly a social news-sharing site, it might have been easier to determine what constituted a good contribution. Comments that were on topic and provided additional information or insight into the material linked could be upvoted with confidence. However, reddit now functions as a community or an aggregation of communities rather than functioning solely as a link aggregator or social news-sharing site. Thus, the guidance that reddiquette provides inadequately addresses diversity of content on the site. For example, what constitutes a "contributing" posting in the numerous advice and social support subreddits is likely to diverge significantly from other subreddits. So what constitutes a meaningful "contribution" to the subreddit /r/aww (which often contains just a bunch of comments discussing how cute the animal pictured is or relating other anecdotes about people's pets—all of which are usually upvoted) is radically different from what is accepted in a subreddit like /r/AcademicPhilosophy, where reasoned debates are expected and encouraged, while empty commentary would be met with derision.

As I mentioned above, reddiquette articulates a logic of individualistic rationality over other ways of being/knowing that might place more value on community harmony or emotional honesty, for example. And, while niche subreddits might encourage these other modes of interaction if a given posting becomes visible to the larger reddit community by being upvoted often enough to reach /r/all, all bets are off. The 2X defaulting debacle demonstrates this—much of the dismay over the subreddit's new status came from subscribers used to a particular form of interaction that was more supportive and often less combative than some other areas of reddit.

Reddit's obsession with and repudiation of its own history

Communities have histories, and they often define themselves through or reassert their boundaries through collective memories. At the same time, too much emphasis on past events may inadvertently drive away new members, as they may perceive the community as "closed" to new ideas and content. Reddit is no different. One ongoing issue is the lack of original content (OC) on the site and the growing suggestion that the site is plagued with reposts.

Given the demographics of many of reddit's users and the sensibilities they share that I detailed in the last chapter, it is probably no surprise that they can be a bit obsessive about their own history. For many redditors, nostalgia about their own childhoods in the 1980s and 1990s becomes a way of connecting to others on the site and creating community. For other redditors, this sense of nostalgia is projected on to the reddit platform and community itself. These are often individuals who have been longtime users of the site and who have seen the various ways in which the community as a whole has changed.

Subreddit "drift"

Redditors are preoccupied with talking about reddit's own past, as evidenced by the ongoing discussion and disagreement over how certain subreddits have evolved over time. The subreddit /r/WTF is a particularly common target for discussions about how the community continues to change. The /r/WTF subreddit most often contains shocking or gross material (often graphic images from medical procedures) meant to elicit a "what the fuck?" response. However, it is a common theme within the comments on posts that are seen as too tame or not shocking enough for reddit members to note how far /r/WTF has strayed from its roots. This is usually followed by a suggestion (often upvoted) that the content is boring or not shocking enough and should be posted instead to /r/mildlyinteresting (a subreddit dedicated to content that is merely "interesting" but not sufficiently provocative for /r/WTF). Within an indeterminate amount of time, the cycle continues, as members submit and upvote more grotesque material to /r/WTF, and commenters praise the subreddit for "getting back to its roots."

Other subreddits regularly experience some version of this phenomenon as well. In early 2013, a post was upvoted to the top page of /r/Music (1218 points) asking what the "point" of the subreddit was, suggesting that posters were mostly concerned about gaining karma rather than posting substantive posts: "The only bright spot of this subreddit that I see is the Fridays where only self posts are allowed, which makes for some very interesting conversations. If I'm missing the goal of this subreddit entirely, then feel free to let me in on it, but as far as I can see, /r/Music just seems like all the karma whores in one spot" (goldfish188, 2012). Interestingly, one of the top-voted and most-replied-to comments read, "Every time I actually bother browsing R/music there is a post like this on the front page." A moderator later commented that the point of /r/Music is to appeal to a large cross section of subscribers, arguing, "If you're coming here to find new, unknown, underground or niche music, you are in the wrong place" and suggesting that other niche subreddits fulfill that purpose. This exchange suggests two things. First is that content posted to certain subreddits may fulfill the subreddit's stated or implied purpose and may even be upvoted but may also be perceived by some portion of the subscriber base as uninteresting, off topic, or otherwise uninspired. Second, this interaction implies that complaints of this manner are frequently voiced among members and upvoted, resulting in a cycle whereby a portion of the subscriber base is frustrated by the tenor of a subreddit, notes their frustration, others post that notes of this frustration are frequent and annoying, and moderators suggest that the frustration is unwarranted given the subreddit's intended purpose. This exchange highlights the double edge of being a popular subreddit. Lots of subscribers ensure a large number of submissions, but these are often reposts (and result in complaints of there being reposts). But subreddits with fewer subscribers tend to have difficulty maintaining a large enough reader/submitter base to remain viable in the long term, while usually maintaining a high level of original content (OC).

Subreddit drift and its resultant complaints are often the result of inadequate or ineffective moderation. Such was the case with /r/atheism right before it was removed from the site's defaults in 2013. Some redditors complained loudly about the increasing amount of "fluff" content on the subreddit: large numbers of memes, irrelevant images of the Milky Way superimposed with aphorisms attributed to Carl Sagan or Neil deGrasse Tyson, and self-posting rants about being an atheist in a sea of theists. By early 2013, moderators forbade submissions that linked directly to memes—memes could be included only in self-posts. Because self-posts earn no karma for the OP, the hope was that this change would decrease the number of memes being submitted to /r/atheism. Some subscribers were unhappy with the change, and cries of "censorship" abounded. More moderator/user drama followed, and finally memes were disallowed altogether. New subreddits with fewer submission rules were created in retaliation (such as /r/atheismrebooted), with the hopes that subscribers would move there. More drama ensued, with someone unloading the philosophical big guns in what would later become a favorite copypasta: "If this subreddit is not open and free, then I honestly don't see the point. Socrates died for this shit and we're taking it too lightly" (skeen, 2013). After more moderator reshuffling and conflict, the subreddit was undefaulted in July 2013 (a complete history of the /r/atheism debacle can be found at UnholyDemigod, 2014). The /r/atheism story is just one of many examples of how poor moderation can contribute to a subreddit's "drift" and eventual dismantling as an effective community.

Repost policing and memetic retelling

As mentioned in the last chapter, reddit can be extremely cynical about itself. One way in which this manifests is through complaints about a perceived lack of original content (OC) in favor of reposts of older content. When posters submit links or content that has been posted before on the site, even if the post gets upvoted, commenters will lament about the fact that reddit constantly recycles its own material and that the original poster is merely "karmawhoring," or sharing content simply to accrue karma (see my discussion about this in the next chapter). This happens despite reddiquette's suggestion not to complain about a repost, as it is likely that not everyone has seen the content before. At the same time, complaints about reposts seem to tie the community together, as they create a kind of collective nostalgia about how the site's content was once fresh and new.

Certain reddit posters seem to take some delight in pointing out that a particular image or meme has already been posted to the community, suggesting that the community is most interested in fresh OC. This occurs despite the fact that many repost threads may even be upvoted to the front page with comments that suggest either (1) not all of reddit has seen the material before and thus it is worth sharing again, or (2) that there is something of value (usually, it is something especially humorous, useful, or shocking) that delights the sensibility of the reddit community. Still, the general tenor of reddit discourse seems to be a lament of OC and a general distrust of the motives of individuals who repost. On occasion, an individual will suggest that reddit's search engine is difficult to use and it may be that the original poster (OP) tried to search the site but was unaware that the material had appeared before. More often than not, however, someone will conduct a reverse image search using karmadecay.com to determine whether content has already been submitted—and report the results publically within the thread (as an example, see alanwins, 2014). Still, it seems that reddit's general sense of skepticism is directed toward the individual who reposts the content, as if he/she is well aware of the material's prior popularity and is recycling it solely for karma points. At one point, for example, the reddit community turned its attention to a particular user (/u/Trapped_In_Reddit), who was accused of reaching the front page of the site solely by scouring the archives, finding top-rated comments on particular postings, and reposting the content (perhaps using a different account) and making the same comment(s) that had been upvoted before on the new post (theempireisalie, 2012).

However, reposting is inevitable for a number of reasons. First, there is the constant stream of new members who may not know that something has been posted recently. Second, and related to the first point, there is the paucity of useful ways to search reddit to see whether something has already been posted. Reddit's search engine is something of an ongoing joke to the community, with many redditors saying they rely on using Google to search reddit as it returns much more reliable and relevant results (antsav888, 2011; girrrrrrrrrrl, 2012). The search functionality on the site is complicated by the fact that redditors often rely on clever titles that contain limited information about the actual content they are posting, as witticisms seem to be one way for a post to gain upvotes. This is particularly true for the image-based content that fills many of the default subreddits; you are far more likely to see a picture of a

Shiba Inu dog with the title "Came home to this face staring at me" than any sort of actual description of what the photo contains. Third, and perhaps most detrimental to the growth and relevance of reddit, is the fact that there is certain content which is likely to be upvoted by redditors—even if it is shared regularly. So certain GIFs will likely be upvoted no matter how many times they appear, especially in the default subreddits. At the same time, complaints about this run rampant on reddit as well—see below for more detail about reddit's "circlejerk."

Complicating matters is reddit's delight in a retelling of certain memes that have circulated within the community. For example, posters will often post the comment "Colby 2012," recalling a particularly dark /r/AskReddit thread in which a poster asked for advice about dealing with his son who he caught sexually molesting the family dog named Colby (concerneddad1965, 2012). Posters responded with supportive comments, advised him to seek professional help, and implored him to make sure the dog was safe. The posting also became a sort of cultural touch point for the reddit community, resulting in referrals to the dog in unrelated threads and becoming an ongoing subject of posts to /r/circlejerk. The phrase "Colby 2012" also points to a larger historical moment: the Kony 2012 short film (http://www.kony2012.com) about Ugandan warlord Joseph Kony that virally spread through Facebook and other social networks in the spring of 2012 (Nahon & Hemsley, 2013). According to Pew's Internet & American Life Project, as of March 2012, 58% of young adults had heard about the video (Rainie, Hitlin, Jurkowitz, Dimock, & Neidorf, 2012). The video itself became a topic of ridicule among most reddit posters, both because its popularity was associated with evangelical Christians (a target of much reddit scorn, especially in /r/atheism) and because it spread specifically through Facebook (which also draws reddit ire). So the "Colby 2012" phrase served a dual function: it was emblematic of reddit's subversive and often perverse style of humor in its reference to the original thread, while at the same time implicitly demeaning the effort of Kony 2012 supporters as misguided or naïve. Interestingly, whenever the phrase was uttered on a posting, it would invariably receive a number of upvotes and replies, mostly along the lines of "Poor Colby" and links back to the original post that appeared on /r/AskReddit. Thus, it seems that reddit members often codify their membership in the community by retelling important stories or events, even (or especially) if they are dark or titillating, while refracting/remixing them through a lens of humor. These exchanges both solidify membership in the reddit community by signaling, "I've been here for a while, and I remember that" while remixing the original meaning and renewing its relevance to the community.

Eternal September and /r/SummerReddit

There is a turning point in reddit's history after which, the story goes, the community changed irrevocably because of a flood of new users who altered the tenor of the conversations and content on the site. Content became more banal, memes more prevalent, and images became the norm in many subreddits. Redditors have referred to this as "Eternal September"—a name originally given to the moment AOL started indexing Usenet message boards and a flood of new users started accessing them (G.F., 2012). Until that point, Usenet had always experienced an influx of new users each September when new college freshmen gained access to the internet for the first time. These new users would be unaware of the norms of interaction in the spaces in which they were engaging and would also have little understanding of the community's shared history. As a result, more senior members would spend a larger percentage of their time moderating these spaces and reminding newbies of the rules of netiquette (G.F., 2012). But Eternal September was different, as it marked the moment that the "masses" discovered the internet, turning a formally exclusive domain populated by early adopters—mostly scientists, educators, programmers, and researchers—into the open space we know it to be today.

So what, exactly, marked the "Eternal September" for reddit? Redditors were posting about the decline as early as 2007, when a poster submitted a Wikipedia article about "Eternal September" to /r/programming with the title "The Eternal September—A Good Analogy for What's Happened to Digg and Is Starting to Happen to Reddit" (jkerwin, 2007). And when comments were first enabled in 2005, community members decried the inevitable influx of memes and off-topic commentary that they believed would occur because of it (Nutshapio, 2005).[4] While perspectives vary, many redditors seem to believe that the migration of Digg users to reddit in August 2010 after a disastrous rollout of a new interface design marked reddit's Eternal September (calmbatman, 2013; ForWhatReason, 2011; Lardinois, 2010; B. Sullivan, 2013).

This is a natural progression in all online communities—they ebb and flow as new users enter into the space, many of whom might question certain naturalized assumptions by breaking norms or by contributing inappropriately or ineffectively early in their membership. However, not all redditors view the migration as necessarily a bad thing. As one redditor who migrated from Digg argued, both sites contained a large group of casual users and a smaller group of "elitist" ones who waxed rhapsodic about the earlier days of the community and decried its changing nature:

> I migrated over from Digg … TBH—Reddit seems to becoming exactly what Digg was: large group of casual users that are pretty chill with a growing minority of elitist users who "yearn for the days of the past."

> I have seen so many mods in the past year banning certain content and breaking up large subs into a billion different subs (I am looking at you /r/android). Just reminds me when Digg creators wanted to change everything and it's what drove the user base away. Some of the users here seemed to be obsessed with "no memes, no pics, no reposts" and millions of other rules. I don't know why Reddit has to be so serious about content—I enjoy both. Laugh at the memes and read the articles. If it's not funny/informative—downvote—move on. If it's a repost—downvote—move on.

> I feel like some users have turned into hipsters and only want to see informative content like Reddit is some PhD convention. Shit is the internet. I like laughing at an FU comic, reading a North Korea AMA, watching Michael Jordan's 50 top plays and then laughing at some gif of a seal dunking … that's Reddit! /rant (whats_hot_DJroomba, 2013)

Some redditors also argue that another form of "Eternal September" takes place every year when US high schools let out for the summer months in early June. "Summer Reddit," as it is called by some redditors, refers to the notion that the site experiences an increasing number of reposts and "shitposts" (discussed in the next chapter), and discussions become even more pedantic than usual because a younger audience infiltrates the community. One of the site's many meta-subreddits, /r/SummerReddit (SR), chronicles reddit's supposed decline in the summer months. The subreddit's sidebar description suggests its purpose is to highlight posts made by "summerfriends" that "highlight a younger viewpoint and lower quality of content." Posts to SR link to content from other subreddits (often the defaults or more popular ones like /r/AdviceAnimals) that are highly upvoted, despite featuring inane or pointless content. Many postings feature a bastardized version of leet or text speak in their titles ("NetLingo List of Chat Acronyms & Text Shorthand," 2014). For example, one top posting, titled "LOL IM SO RANDUM CHOOCHOO" linked to a recent /r/AskReddit thread in which someone asked what the worst question anyone could think of to post to ask reddit would be (razorbeamz, 2014). Shockingly, the insipid self-post made it to reddit's front page (with 1105 points as of this writing). Often SR showcases the worst of reddit's tendency toward hipster or "edgy" racism or sexism. For example, one posting linked to a /r/funny stock photo of a black male scientist holding up a test tube with the words, "Finally Watermelonium" in the title—the SR posting featured the title, "BLACK PEOPLE LIEK WATERMELONS GUYS EDGY AS FUCK" (OverjoyedMuffin, 2014).

Whether or not "Summer Reddit" is a real phenomenon is debatable. The idea of a drop in quality on reddit is a meme likely borrowed from 4chan, whose users claim an increase in "Summer Fags" every June—this despite founder Christopher "Moot" Poole's insistence that it is not actually true ("During summer, is there actually an increase," 2013). A posting to /r/TheoryOfReddit suggests that one kind of offending content—reposts—is actually most common in September when high school and university students are back in school ("During summer, is there actually an increase," 2013). Still, the perception that content on the site decreases in quality during the summer becomes a point of discussion every year between May and September, with comments like "damn summer reddit" regularly appearing on poor content. However, it is most likely that the "Summer Reddit" effect is an issue of selection bias, with redditors paying more attention to the perceived drop in quality during summer, when in fact these kinds of terrible postings are always present on the site.

Meta-subreddits

Toward the end of my fieldwork, I changed my browsing habits considerably. At the beginning of the project, I scanned /r/all for hours, clicking on links, taking copious notes, and occasionally commenting or posting to my subscribed subreddits. I would chuckle here or there or groan in frustration or find myself sending out random tidbits of internet culture that I had found on reddit to my friends and family. However, my focus shifted as I started to get a better sense of reddit culture. I started predicting with some accuracy what the most upvoted comments would be on a given submission or what popular culture reference would be made or how the conversation would derail into other, only peripherally related topics. Soon, I found myself growing tired—of the project, of reddit, and of screens in general. Such is the life of the digital ethnographer: your field site is a screen, your work is a screen, and your life is, in essence, a screen. This is probably a predictable cycle within any research endeavor; the excitement you feel in the beginning inevitably changes to a sense of tedium and drudgery toward the middle and end of the project no matter how passionate you are about the topic. As a distraction, I found myself revisiting subreddits that I had only cursorily examined in my earliest reddit days—smaller ones that had not made much sense to me when starting the project as they all discussed reddit from an insider perspective. These could all be classified in the category of meta-subreddits: subreddits dedicated to engaging in meta-talk about reddit. Some focused on more "academic" examinations of reddit (like /r/TheoryOfReddit and /r/circlebroke); others pointed to "drama" happening in various subreddits and pulled out the metaphorical popcorn to watch it unfold (/r/SubredditDrama and /r/Drama); and still others served as a place to get caught up in reddit lingo or past happenings (/r/OutOfTheLoop and /r/MuseumOfReddit). While I had visited /r/TheoryOfReddit when I was first designing this project, I admit that many of the meta-subreddits were baffling to me as a relative newcomer to the community. But toward the end of my fieldwork, I found them invaluable resources for both challenging *and* confirming my observations about reddit culture. In addition, they offered enormous relief from the tedium of interactions I saw unfolding on the default subreddits filling /r/all. I also found them (along with /r/ShitRedditSays, a unique meta-subreddit which I discuss in Chapter 6) a welcome space of mostly like-minded individuals, who were often as annoyed and disgusted as I was by the casual racism, sexism, and homophobia in which many redditors seemed to revel.

Meta-subreddits provide further evidence of reddit's obsession with examining its own present, past, and collective sense of "self." By engaging in meta-talk about reddit on reddit, meta-subreddits represent a unique and, I would argue, critical aspect of reddit culture—one that seems to be missing in many other participatory culture platforms and social networking sites. Although not all meta-talk about reddit occurs only in the meta-subreddits—as subscribers to /r/TheoryOfReddit, for example, are also subscribed to and participate in myriad other subreddits—meta-subreddits exist specifically for the examination of reddit's unique patterns of interaction and platform logics.

So why does the community encourage such a rich culture of inward focus and interest in analyzing the minutiae of reddit? I can think of several reasons. Much of it is likely driven by the tone reddit administrators and founders set: emphasizing reddit as a space for information sharing and niche interests and encouraging a sense of community ownership. The platform's open-source nature certainly contributes to this as well. In addition, the reliance on a large team of moderators to manage subreddits, adjudicate disputes, and generally keep the site filled with new and interesting content means the community is invested in the direction of the platform's development. As one redditor commented in a /r/TheoryofReddit posting, moderators often engage in meta-conversations about reddit to better understand its nuances and how they might improve their own subreddits:

> In some ways, reddit is like a little nation. it's got it's own over-arching culture, but there are individual states (subreddits) that have their own cultures as well, and the way those interact can be fascinating some times. Some are huge and have great influence, some are smaller, perhaps more secluded.
>
> Many of us also happen to be moderators, so it's beneficial for us to understand the communities we manage as much as possible. (Addyct, 2014)

Conversations about reddit are likely to happen on reddit, and administrators actively encourage these kinds of discussions by posting updates about new community initiatives, changes to the default subreddits, and tweaks made to the codebase to the reddit blog. Founder Alexis Ohanian (2013) has noted the critical importance of listening to and engaging with the community, lest the community go elsewhere. These meta-conversations also reflect an interest that many redditors seem to express in understanding how things work and how to improve them. This likely stems from interest in STEM fields, where "design thinking" and "engineering thinking" both encourage this kind of orientation (Buchanan, 1998; Dym, Agogino, Eris, Frey, & Leifer, 2005). Discussions on /r/TheoryOfReddit (TOR), for example, often ask questions or pose hypotheses about why a particular thing on reddit works the way it does or how it could be improved. As of this writing (late June 2014), some of the postings on the front of TOR include questions asking subscribers how much their reddit account reflected their personality (goldguy81, 2014), whether reddit is increasingly becoming a space for hateful ideas (Pekhota, 2014), and why some accounts seemed to be "singled out" for celebrity status (Ooer, 2014). Last, I believe it has much to do with the way that the reddit community views itself as somehow exceptional or "special" (which also manifests in its reactions to outsiders, as mentioned in the last chapter)—and that this exceptionalness or specialness must be examined, criticized, quantified, and/or improved.

At the same time, meta-subreddits have a mixed reputation. They are often framed as being elitist and overly cynical about reddit. Some subreddits, such as /r/Polandball (dedicated to sharing crudely drawn comics playing on stereotypes of different countries) and /r/TwoXChromosomes, forbid outside linking—perhaps because they fear the possibility of vote brigading or that redditors from /r/SubredditDrama (SRD) will start conflicts with subscribers in their subreddit and then link to it as an example of drama. So they ban redditors from posting, which is more symbolic than anything else, as new accounts can be created easily. But it is true that commentary in the meta-subreddits like SRD can get a bit snarky—discussing the "rest of reddit" as if they were alien creatures. But as with other controversial subreddits (such as /r/ShitRedditSays, discussed in Chapter 6), meta-subreddit subscribers are still an important and valid part of the reddit community—even if they are making light of, parodying, critiquing, or otherwise analyzing other subreddits and their members.

Remember the karma while tipping your fedora: /r/circlejerk

Probably the most prominent and notable meta-subreddit is /r/circlejerk. It is the main hub of a large number of subreddits that look inward and parody an aspect of reddit culture or a specific subreddit—from /r/moderatorjerk (which lampoons moderator interactions with subscribers) to /r/Gamingcirclejerk (which parodies the discussions that occur on /r/gaming).[5] With a bit over 200,000 subscribers, /r/circlejerk is also one of the more popular subreddits (outside of the defaults), and postings frequently make their way onto the front page of /r/all. For novice redditors, /r/circlejerk is a bizarre collection of inane postings, featuring neologisms that are curious portmanteaus of phrases you might see elsewhere on reddit (upmason, downunidans, May May Man, etc.), stereotypes of redditors as fedora-wearing, sexually frustrated meme aficionados who say things like "M'Lady" nonironically. If reddit is a carnival, /r/circlejerk is a pack of carnies mocking the rubes late at night.

Moderators of /r/circlejerk regularly change the subreddit's design in response to current events or to lampoon aspects of reddit culture. For example, when /r/AdviceAnimals banned the "Unpopular Opinion Puffin" (UOP) image macro, /r/circlejerk suddenly changed to look exactly like /r/AdviceAnimals except it featured only images of UOP (see Figure 4.1). They also disallowed any posting that did not feature the puffin in all its glory.

/r/circlejerk's puffins parodied subscribers' unhappiness about the UOP's removal from /r/AdviceAnimals, despite their frequent use as a not-so-veiled attempt at expressing racist sentiments. One UOP on /r/circlejerk went so far as to feature the puffin with a Klansman hood Photoshopped on it. Similarly, after technology journalist Ryan Block leaked a phone call with a Comcast

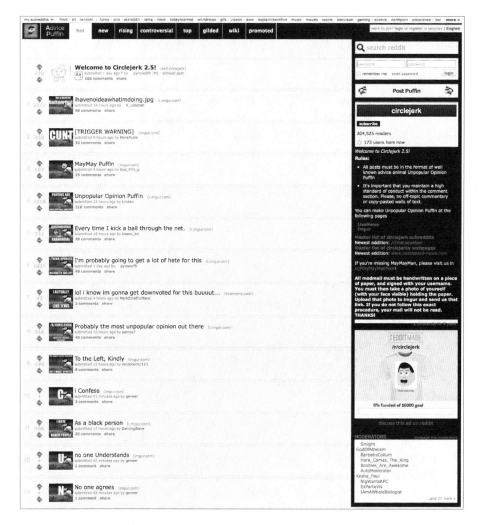

Figure 4.1. /r/Circlejerk's response to the banning of Unpopular Opinion Puffin (captured May 28, 2014).

representative in which the representative would not allow him to cancel his service (Weissmann, 2014), /r/circlejerk's logo satirized the Comcast logo, changing it to read "Concast" to represent how up in arms some of the reddit community became after the call became public.

No aspect of reddit culture is immune from the /r/circlejerk treatment. A classic example of this resulted from a posting that cupcake1713 (Alex Angel, reddit's Community Manager) made to the reddit blog, reminding site members to have empathy for their fellow redditors in a posting titled "remember the human":

> Hi reddit. cupcake here.
>
> I wanted to bring up an important reminder about how folks interact with each other online. It is not a problem that exists solely on reddit, but rather the internet as a whole. The internet is a wonderful tool for interacting with people from all walks of life, but the anonymity it can afford can make it easy to forget that really, on the other end of the screens and keyboards, we're all just people. Living, breathing, people who have lives and goals and fears, have favorite TV shows and books and methods for breeding Pokemon, and each and every last one of us has opinions.
>
> …
>
> So I ask you, the next time a user picks a fight with you, or you get the urge to harass another user because of something they typed on a keyboard, please … remember the human. (cupcake1713, 2014b)

The post was well received by the reddit community, with redditors offering strategies to "maintain their cool" when interacting on the site and just generally agreeing with cupcake1713's advice. Within hours, a posting to /r/circlejerk appeared, titled "remember the karma," offering a pitch-perfect parody of cupcake1713's original post:

> Hi reddit. dhamster here.
>
> I wanted to bring up an important reminder about how folks interact with each other online. It is not a problem that exists solely on reddit, but rather the internet as a whole. The internet is a wonderful tool for interacting with people from all walks of life, but the anonymity it can afford can make it easy to forget that really, on the other end of the screens and keyboards, we're all just looking for karma. Virtual, meaningless internet points to aspire for in place of lives and goals and fears, earned by agreeing with our favorite TV shows and books and methods for breeding Pokemon (Firefly, Game of Thrones and the Masuda method, respectively), and each and every last one of us has opinions.
>
> …
>
> So I ask you, the next time a user picks a fight with you, or you get the urge to harass another user because of something they typed on a keyboard, please … remember the karma. (dhamster, 2014)

While masquerading as a bit of trivial /r/circlejerk humor, dhamster's posting actually serves as a critique of reddit's preoccupation with karma and the echo chamber that often results from redditors parroting what is popular with the community rather than engaging in meaningful discourse. It also highlights the tendency of the community to argue over the minutiae of relatively inconsequential topics related to popular culture and fandom—and to be impressively dramatic over small differences of opinion. Additionally, it serves as a not-so-veiled critique of the "be excellent to each other" rhetoric into which reddit administrators occasionally lapse.

Watching drama unfold in /r/SubredditDrama

Another favorite meta-subreddit redditors love to hate is /r/SubredditDrama (SRD). SRD chronicles and comments on "drama" that occurs on other subreddits. Drama is, as Marwick and boyd define it, about interpersonal conflict played out in publically networked spaces (Marwick & boyd, 2011a). However, unlike the drama that often occurs among teens on social networking sites, which was the focus of their work, reddit drama is for the most part *confined* to reddit. Perhaps because of the pseudoanonymity of the reddit site and the tight controls the administrators enforce around private information, there are fewer opportunities for the drama to "spill out" to other platforms. Like other spaces, however, drama on reddit is complicated by the tendency for some individuals to troll (Bergstrom, 2011; Phillips, 2015) certain threads or subreddits for seemingly no reason other than to create conflict. For example, a troll might offer up an opinion that consoles are actually awesome in /r/pcmasterrace (a subreddit intent on extolling the virtues of gaming PCs over consoles). Drama, however, differs from trolling in that it results from opposing viewpoints offered in presumably good faith. True drama moves beyond one or more redditors merely sharing contradictory opinions to interactions that turn personal (some variation of "You are the most inane person on the planet," using much more offensive language) or intense discussions that hinge on differences of opinion over minute details, resulting in often nonsensical insults being traded.

Drama on SRD is often referred to as "popcorn" to emphasize its public, performative nature—and the implicit importance of these conflicts playing out in front of a group of spectators, without whom there would be no "drama." Spectatorship as a concept has its roots in film theory but can usefully be applied to mediated spaces of all kinds. As Michele White (2006) argues,

"Spectatorship indicates the processes of watching and listening, identification with characters and images, the various values with which viewing is invested, and how these ideas continue even after the spectator has stopped viewing" (p. 6). Many of the conversations that occur on SRD are about parsing the various positions being taken and, especially, noting the ridiculousness of some of the claims made. Like many other SRD denizens, I enjoy the gossipy nature of the threads that the subreddit chronicles.

Observing drama without being a part of it directly is like watching an episode of *Gossip Girl* or some other guilty pleasure. It is also problematic, in that the act of spectating likely encourages more drama. As is the case with other meta-subreddits, SRD requires a kind of "hands-off" approach with regard to the interactions that happen outside SRD. Actively participating in the drama that is being chronicled is called "popcorn pissing" and is a bannable offense. Only "np" (non-participation) links are allowed to other subreddits to avoid accusations of vote brigading by SRD. And SRD subscribers are encouraged not to bring the drama back from the threads on which they are commenting. However, because a bot posts in the original thread with a link to SRD, individuals associated with the original drama will sometimes show up in SRD to discuss it further. There is even another subreddit for drama that occurs on SRD called /r/SubredditDramaDrama.

While there are the standard dramafests on SRD—race, gender, and guns—much of what serves as the "butteriest" popcorn for SRD is that which occurs on the site's niche subreddits and/or that which involves extremely small-stakes issues. For example, when a redditor posted an album of pizzas he/she had made over the years to /r/Pizza and asked for feedback, several redditors pointed out that said pizzas looked overcooked. The OP disagreed and melted down, with a statement that some on SRD pointed out read like a fine-tuned copypasta:

> I am very secure about my pizza. You have obviously never tasted pizza if you think mine look bad, and that is very sad for you. You should get a pizza right away so you will have some knowledge of pizza and not sound like a fool when you speak. Again, I would ask you to prove your own pizza making skills if you think you are capable of judging others so harshly. But that would require something well beyond your capabilities. You really should try pizza, it is quite delicious and you won't sound so stupid when you comment. (genius-bar, 2014)

To an outsider, such intense preoccupation with banal things like pizza might prove puzzling, but redditors usually hold strong opinions about most things and are not shy about sharing them with others.

From patterned interaction to the reddit circlejerk/hivemind

What makes reddit supremely effective at mobilizing around political issues (as mentioned in the last chapter) can also create a space in which alternate viewpoints are discarded or where deliberation around important public issues happens infrequently.[6] At some point, the patterned interactions that make community possible on reddit also turn it into a kind of echo chamber, where ritualized behaviors become almost empty signifiers with participants unaware of the origins or purpose of the ritual in which they are taking part. Redditors are obsessed with the idea of a reddit hivemind, also known as the "reddit echo chamber" or "circlejerk." Observations of and complaints about the reddit hivemind manifest in numerous ways—and are frequently mentioned and dissected. See, for example, an extensive compilation of most-hated overused phrases offered by the community itself in an extremely meta /r/AskReddit posting, which included: "This," "Found this gem …," "mind blown or mind=blown," "… now KISS," "like a BOSS," "I'm cutting onions," "I have the weirdest boner right now," and "Boy, that escalated quickly," among many, many others (mikey_mike24, 2013). The fact that these kinds of phrases are still upvoted enough to be regularly visible on the site presents a conundrum. My guess is that they are easy ways for novice redditors to feel a part of the community, which can be an overwhelming place given the often quick wit and playfulness that seem to roll off the keyboards of more seasoned redditors. These neologisms both signal membership and convey a kind of reddit expertise. (A number of these phrases and acronyms are listed in the glossary at the end of this book.) The reddit "hivemind" also contributes to a sense of cynicism that some redditors may feel regarding the community. This is where meta-subreddits step in—serving what I would argue is a critical function for more seasoned redditors to engage in meta-discourse around the ins and outs of the site, as well as revel in its drama, or commune with other redditors who may find aspects of the community distasteful and problematic.

Notes

1. Further complicating matters was the fact that reddit administrators stepped in during one of these moderator–subscriber discussions to note that some subscribers were actually lying about the harassment they were receiving through PMs (Deimorz, 2014a).

2. At least this idea that reddiquette is community authored is stated whenever it is discussed by site administrators. Looking at the change history, I noticed that (a) very few public addendums have been made in the past year (only 11, mostly minor changes) and (b) the actual changes to the wiki are limited to high-level administrators (such as cupcake1713, who is a community manager for reddit). This suggests that while reddiquette may reflect community input, it is not actually open for revision by the community as, for example, a Wikipedia article might be. One of my informants noted this specifically, suggesting that the idea that reddiquette is authored by the "community" is a myth.
3. Interestingly, /r/politics does not specify disallowing racist or sexist usernames, as, for example, /r/TheoryOfReddit does. Presumably, a redditor could create an extremely offensive username and still participate freely in /r/politics, which is troubling given its stated policy.
4. This thread is still updated occasionally by redditors, after it was linked to in a posting four years later (circa 2009). In particular, 2009 redditors responded to a 2005 comment that argued that reddit was turning into Digg. Most expressed the sentiment "man, you should see it now," and opened up a new conversation about how the community might be preserved. Some suggested prodigious use of the downvote button, while others argued that it was just an inevitable decline as the site attracted an influx of new users.
5. A list of circlejerk subreddits can be found at http://www.sobrave.net/subs. There is even a list of "geometry circlejerks" (/r/trianglejerk, /r/polygonjerk, /r/hypercubejerk, etc.), which again suggests the almost limitless creativity and ridiculous lengths to which some redditors will go in parodying the site.
6. Although in my opinion, reddit is much more likely to engender deliberation around important topics in a way that other social networks, participatory culture platforms, and even other communities may not. This might be because of the audience it attracts (redditors like to think of themselves as rational, well-informed individuals), it might be partially enabled by the site's pseudoanonymity, and it might also be the result of the large number of moderators that keep conversations on topic per the subreddit in which they are occurring. All of this merits future exploration.

· 5 ·

PLAYING SERIOUSLY

Trying to pinpoint exactly when or why I first visited reddit is difficult. I know I had heard about it through *BoingBoing* and other technology-minded blogs, and it's likely that a number of my friends had sent me links from the site. But I resisted creating an account for as long as possible, despite visiting reddit with some regularity. Why? Well, I was overwhelmed by the sheer volume of content on the site, and, more important, I was awed by the cleverness with which redditors interacted with one another. These people were quick—quick with a pun, a bon mot, a retort, a pop culture reference—and it was intimidating to think about joining in with them. This sense of frivolity and play is part of what I find most appealing about reddit. And reddit is at its best when play serves to offer incisive commentary about those in power—and at its worst when it focuses its sights on those already marginalized or disempowered.

Patterns of play

Despite the extensive level of customization possible on reddit in terms of the subreddits and subcultures to which a redditor might subscribe, there are actually a number of distinct patterns that cross these boundaries and characterize play on the site. Some of these are familiar to many other participatory culture platforms and social networking sites; other patterns of play are shaped more directly by reddit's design and underlying technological logic.

Memes

The most common form of play on reddit and, I would argue, sometimes one of the least interesting, involves the posting of memes, often in the form of pictures, animated GIFs, videos, or modified LOLcats/image macros (for more discussion of memes and their cultural significance, see Davison, 2012; Milner, 2012; Shifman, 2013, 2014). Many popular subreddits, such as /r/AdviceAnimals, consist entirely of conversations inspired by user-created memes. At the same time, discussions on reddit frequently decry the overuse of memes on the site as detracting from more serious interactions, and many subreddits actively discourage or ban memes from posts and threads. Still, given the nature of material submitted to the default subreddits, first timers or those visitors to reddit who choose not to create an account will likely perceive the community as merely a place to exchange silly (or disturbing) memes.

Popular memes on reddit tend to demonstrate many of the same themes that Limor Shifman (2014) has identified in her research into memetic YouTube clips: they involve ordinary individuals, play on images of flawed masculinity, employ whimsical humor, and are both simple and repetitive. Often, these popular memes are remixed with other content appearing at the same time on the site—usually especially humorous, gross, or outrageous material—creating a kind of meme-mashup. For example, two popular memes, a recipe for "2 AM chili" (tylercap, 2011) and instructions for creating a shower using frozen soap and water (EffinD, 2011), appeared in August 2011. Later, these were mashed up into a playful 2 AM ice-chili shower meme, which consisted of a series of pictures of someone freezing a cup of chili, rubbing it on his arm, and having a dog lick it off (berbertron, 2011). As mentioned in the last chapter, re-invoking these memes in later threads becomes a kind of cultural "touchstone" for the community. These exchanges both solidify membership in the community and remix the original meaning and renew its relevance to the community.

My informants suggested that what they perceived as an overabundance of memes on the default subreddits often inspired them to create a curated list of subreddits to reduce the "noise." Additionally, my observations of the site over time suggest a common pattern, whereby popular posts in nondefault, highly moderated subreddits (such as /r/science or /r/AskHistorians) that specifically ban memes experience a greater amount of this kind of off-topic, meme-based conversation if a particular post becomes popular enough to show up on /r/all. And, while such memetic humor is common on the site, it is not particularly unique to the space, nor do memes demonstrate the most creative examples of the community's approach to play.

Pun and pile-on threads

Pun threads are near ubiquitous on reddit. They appear in response to many postings, particularly those in /r/pics or /r/funny. The community seems to be divided on this behavior: while many of the threads are upvoted and feature contributions from many different members of reddit, other posters seem merely to tolerate or may actually deride the prevalence of these threads in their own comments. Still, redditors seem to appreciate a clever bon mot. An example: a redditor posted a picture of a rock to /r/pics that had split open on the construction site on which he/she was working (flobbley, 2013). A highly upvoted response was a serious one from a geologist identifying the rock as slag. Another redditor humorously responded, "That's not a nice thing to say about a rock. I'm sure it's of good moral fiber." And someone else posted, "Not very gneiss at all. Cleavage." The thread continued with others making puns using the names of various rock types. The thread then returned to more serious matters with someone else asking for more information about slag and getting an accurate response about its origins as an ore by-product, to which others responded humorously, suggesting that it is actually the by-product of Unobtainium (a reference to the movie *Avatar*) or Adamantium (a reference to the Marvel Comics universe). References are also made to *Game of Thrones*, *The Wire*, *Breaking Bad*, and *Borderlands 2*. Actually participating in these sorts of interactions—or even just understanding what's going on and why the responses are intended to be read as humorous—requires a dizzyingly large familiarity and facility with US popular culture from movies to games to comic books to television shows and beyond. It also connects many of the conversations that happen on reddit to the notion of a kind of "geek masculinity" more broadly, in which passion for understanding and discussing the minutiae of a particular domain is pleasurable and serves important social functions within these communities (Burrill, 2008; Coleman, 2013; Taylor, 2012).

Pile-on threads take several different forms. One common pattern is for someone to make a pop culture reference in response to a post and then others fill in a line of dialogue or make another reference to the same fictional universe. For example, images of ocelots or even cats that look remotely like ocelots posted to /r/aww or /r/pics will inevitably lead to someone making a reference to the animated adult television show *Archer*, in which the main character interacts with an ocelot (see, for example, buckysbitch, 2013). Comments repeating phrases from the show's dialog such as, "He remembers me!" "Look at his tufted ears!" and "You should get him some toys. It's like Meowschwitz in there!" are posted as one-liners by individual accounts. Or if someone mentions ice cubes, another redditor might respond with lyrics from Vanilla Ice's "Ice, Ice, Baby" (AyChihuahua, 2012).

A second version of the pile-on thread involves individuals remixing and playing off someone else's posting by modifying the original object in some way. For example, a thread titled "The Saddest Cookbook" linked to a cover picture for *Microwave Cooking for 1* (JohnArr, 2013). A redditor posted a response, with the comment "I made it less sad," and linked to an edited version of the cover with another line added to the book's title, making it read, *Microwave Cooking for 1 Awesome Party*. Another commenter included both the original title and the modification with a comment, "Back to reality" and linked to another version of the cover with the words "where you are the only guest." The thread continued, with different individuals adding more words to the book's original title or modifying the cover in some other way in reaction to what others before them had written. (See Figure 5.1.)

These pile-on threads are reminiscent of the Surrealist game *Exquisite Corpse* (Flanagan, 2009), in which individual wit and clever commentary come together to create a collective collage that is greater than the sum of its parts.

Reaction GIFs

Reaction GIFs, or looping animated images that respond to a posting using little or no text, are a popular form of play in many subreddits. In her study of reaction GIFs used on Tumblr, Katherine Brown (2012) argues that they serve "interpersonal, emotive, [and] humorous" functions (p. 52). While GIFs are not limited solely to this space, as Tumblr, 4chan, and message board users also engage in GIF-based humor, reddit is largely known for its ability to generate and distribute reaction GIFs. It was reddit's /r/gifs community to which Jason Eppink, the Associate Curator for Digital Media at the Museum of the Moving Image, reached out when creating an exhibit on GIFs—asking redditors to

Figure 5.1. Example of a pile-on thread—"The Saddest Cookbook."

pick the reaction GIFs they viewed as "classic" and describe what they meant (jasoneppink, 2014). As the exhibition notes suggest,

> The reaction GIF has emerged as a form for communicating with short moving images in response to, and often in lieu of, text in online forums and comment threads. These animated GIFs consist of brief loops of bodies in motion, primarily excerpted from recognizable pop culture moments, and are used to express common ideas and emotions. Understood as gestures, they can communicate more nuance and concision than their verbal translations. While many reaction GIFs are created, deployed, and rarely seen again, some have entered a common lexicon after being regularly reposted in online communities. ("Museum of the Moving Image," 2014)

The exhibition included 37 different reaction GIFs suggested by redditors.

Like image macros and other memes, the effective reaction GIF is emotionally evocative, often humorous, and precise while still being broad enough to apply to a number of different situations. As with other content on reddit, those GIFs that are less frequently used in threads or in popular subreddits are usually received with more interest and upvotes. Still, a reaction GIF's effectiveness lives and dies by its ability to encapsulate a specific response creatively and precisely while still expressing a kind of universal sentiment with which others can identify. Within threads, a perfectly timed reaction GIF can yield a string of other reaction GIFs, creating a kind of playful call-and-response. Popular reaction GIFs, particularly those that include some sort of text overlay, are often shorthanded. For example, a GIF of Sweet Brown saying, "Ain't nobody got time for that" on a news broadcast after her home caught fire is often written out as "aintnobodygottime.gif" (or a variation thereof) instead of linking to the actual GIF (see ColtenW, 2012, for more about this meme). Since the Sweet Brown reaction GIF has been posted numerous times to reddit, it can be shorthanded in this way and still retain its memetic context. These kinds of interactions depend on members retaining a fairly long and robust sense of the community's history, as discussed in the last chapter.

Several subreddits exist solely as spaces for individuals to post and comment on different reaction GIFs. These include /r/reactiongifs (a large, catchall subreddit for these kinds of images), /r/catreactiongifs (reaction GIFs featuring cats), /r/analogygifs (images that can be used to encapsulate a particular experience), and /r/TrollXChromosomes (reaction GIFs and other memes from a female perspective). There are also larger catchall GIF subreddits such as /r/gifs and /r/HighQualityGifs that include all sorts of GIFs that are not "reaction GIFs" per se. Postings to reaction GIF subreddits typically start with some sort of acronym, such as "HFW" (How I Feel When …) or MRW ("My Reaction When …") with the GIF standing in for the way the OP felt or responded to a given scenario detailed in the posting's title. One reaction GIF may be recycled throughout the site in multiple ways: perhaps appearing as OC (original content) in /r/gifs, then being posted to /r/reactiongifs or /r/TrollXChromosomes with an "MRW" and brief story, and finally appearing in a thread on /r/funny in response to someone else's comment without any context required to understand its meaning.

A second type of reaction GIF is specific to reddit. This is the upvote/downvote reaction GIF, which indicates how a redditor reacted to a given post. Often these reaction GIFs are remixed versions of other GIFs, with either an upvote or downvote arrow graphic replacing a physical object from the original GIF. A paradigmatic example is an upvoting reaction GIF borrowed from the movie *American Psycho*. In this GIF, Christian Bale's character Patrick Bateman throws a business card at someone. The remixed, reddit-specific reaction GIF replaces the card with an upvote arrow. This kind of reaction GIF would appear in a thread when someone wished to indicate that he/she was upvoting the prior comment or posting as a whole. Likewise, there are downvote reaction GIFs that indicate displeasure, disagreement, or some other motivation for downvoting material. Still others are remixed in such a fashion that a downvote might become an upvote, or vice versa. Like other reaction GIFs, there are several subreddits for sharing upvote/downvote GIFs, the most popular of which is /r/upvotegifs.

As with novelty accounts, which I discuss below, both site members and moderators have mixed feelings about reaction GIFs outside of subreddits dedicated to sharing them. On the whole, redditors enjoy image-based content, as it is quick to digest, requires almost no effort to load (assuming one has Reddit Enhancement Suite installed), and is regularly upvoted. One posting to /r/TheoryofReddit even argued that reddit should no longer be considered a link aggregator; instead, it should be seen as an image board—especially for those who are not logged in or only subscribed to the defaults. According to one redditor, an assessment of a single day's top 200 posts indicated that 70% of submissions were images (Maxion, 2013). Like other image content, reaction GIFs are both upvoted regularly and quickly become stale—with redditors complaining about threads being derailed by pointless reaction GIFs while at the same time upvoting them so that they remain visible. /r/circlebroke discussed this phenomenon in a September 2012 posting, with commenters noting that although a reaction GIF used sparingly can elicit laughter or otherwise perfectly encapsulate a feeling in a portable nugget of animated goodness, it can be a sign of laziness and can tend to run any meaningful discussion (even a playful, humorous one) into the ground (BlackbeltJones, 2012). As one commenter argued, "I really hate seeing .gifs involving clapping or thumbs up-like gestures because they are the epitome of worthless effortless karma whoring bullshit. And the worst part is that people will reply with similar .gifs to the original gif and create a whole comment chain of shit that contributes nothing to the discussion."

Novelty accounts

As discussed earlier, the reality of reddit as a pseudoanonymous space (Donath, 1999; Hogan, 2013) stands in stark contrast to the notion of a unitary identity championed by social networking sites such as Facebook or Google+. Given this fact, along with the large number of accounts present on the site, the low barrier to entry for registration, and the interest in creating a unique, memorable username, a pattern of play relatively unique to reddit has emerged: the novelty account. The username of a novelty account often indicates the kind of content it posts. Some are mundane, albeit humorous attempts at gaining karma—for example, /u/JudgeWhoAllowsThings posts mostly one-line quips such as "I'll allow it" in response to others. Others demonstrate creativity by responding to threads using something other than simply words: /u/ICanLegoThat responds to posts with pictures of Lego scenes he has created, and /u/SingsYourComment links to a singing, rather than written, response.

Still others provide an even more quirky, humorous flavor to interactions on the site. For example, /u/DiscussionQuestions responds to postings in the style of discussion questions one might hear in a high school English course. Posts by /u/GradualBillCosby start out seeming to provide a serious response, but soon descend into absurd discussions of the merits of Jell-O pudding (Bill Cosby was a longtime spokesperson for the food product in the US). Two of the more popular novelty accounts, /u/Shitty_Watercolour and /u/AWildSketchAppears, respond to threads by posting watercolor images and pencil sketches, respectively.

Some novelty accounts exist solely to offer a clever response or further a joke in a thread. For example, in a thread in which someone mentions "Ned the Cosmic Manatee" (a roundabout reference to the teddy bear from the movie *Ted*), a user might create such an account as a humorous response. In this case, a redditor created a new account (/u/NedTheCosmicManatee) just to post to the thread, and received a large number of upvotes as a result (djmushroom, 2014). Accounts that have been around longer, that is, accounts that are not created on the fly for a specific interaction, are met with less skepticism. Redditors will often respond with something like, "Account for 8 months. I'll allow it" to suggest that this kind of humor is of a higher quality because it took more effort than simply creating an account at that moment for the interaction. This kind of play is prevalent on reddit, especially in the more heavily trafficked default subreddits, as they both tend to have looser moderation rules and many more comments than a thread in smaller subreddits.

However, not all novelty accounts are welcome additions to the site's discourse. /u/PigLatinsYourComment, which, as the name suggests, posted an exact replica of a comment translated into Pig Latin, was downvoted regularly, with others responding unfavorably and suggesting that the account's owner must be trolling. This suggests that there are certain unspoken rules around when and how a novelty account is welcome to comment on a post. Not surprisingly, my informants had a conflicted relationship with novelty accounts. Some viewed them as nuisances as they offered little in the way of meaningful comments, but others enjoyed stumbling on them—as interviewee Jessie offers, "… [they are] the fairies of reddit. You don't know when they're coming, but when they do they bring a gift for everyone to enjoy!"

Bots

As with other participatory culture platforms, most notably Twitter (Mowbray, 2014), reddit features a number of automated scripts written by redditors called "bots." These bots populate the site and interact with redditors in various capacities, engaging in what Taina Bucher (2014) argues could be termed *botness*, or "the belief that bots possess distinct personalities or personas that are specific to algorithms." Unlike the bots and novelty accounts that populate Twitter, however, bots on reddit are not "followed" purposefully—they are, instead, stumbled upon serendipitously when someone is browsing a comment thread. Thus, the experience of encountering a bot is much like the way in which my informant mentioned encountering a novelty account—that it is a (sometimes welcome) surprise.

Reddit bots serve multiple functions on the site. Some of them are created for purely pragmatic purposes, such as /u/CaptionBot, which transcribes the visual text on image macros in case a hosting site is unavailable or blocked. If /u/Jiffybot is mentioned in a posting that includes a time-stamped link to a

YouTube video, it will automatically transcribe that portion of the video into an animated GIF. Other bots manage daily, low-level moderation tasks, such as blocking links to a particular site that has been banned by a subreddit's moderators or indicating that a posting has been linked to in another subreddit, as is the case with /u/totes_meta_bot. Some bots, such as /u/redditbots, take snapshots of threads for archiving purposes in case the original posting or comments in the thread are deleted by its participants. Still others provide additional context about facts or figures posted to the site. For example, /u/MetricConversionBot offers conversions between imperial and metric measurements in comment threads. If summoned, /u/CompileBot executes code (C, Java, JavaScript, PHP, Perl, etc.) in reddit threads (SeaCowVengeance, 2014). The bots /u/autowikibot and /u/Wiki_Bot automatically post the first paragraph of a given Wikipedia article as a reply if a redditor includes a link to that article in his/her comment. The bot /u/astro-bot posts coordinates and an annotated image if a particular star is mentioned. The bot /u/bitofnewsbot creates an automated summary of a linked news article posted to /r/worldnews and posts it as a comment ("About Bit of News," 2014). The bot /u/VerseBot posts the text of whatever biblical verse is mentioned. And the bot /u/gandhi_spell_bot shows up whenever a redditor misspells Gandhi as "Ghandi" and provides a correction.

Others provide a kind of automated social support. For example, /u/Extra_Cheer_Bot offered an upvote whenever someone used the terms "I am sad" or "depressed." However, this bot was later banned from a number of subreddits, including /r/depression, and finally deleted from reddit altogether two weeks after its implementation, as its message could trigger or otherwise prevent individuals from getting the real social support they needed (laptopdude90, 2014). It was rebooted and found new life as /u/BeHappyBot. BeHappyBot appears in threads where someone appears unhappy and offers a virtual hug with the following text: "Hi there. It seems you're sad. I can't tell if you're messing around or you're serious, but if you need someone to talk to, my master is always available for a chat. Either way, I hope you feel better soon! Have a hug! (つ'3')つ."

Still other bots engage in a kind of automated play. These include accounts such as /u/haiku_robot, which automatically reposts any comment that fits the standard five-seven-five-syllable haiku structure. /u/Mr_Vladimir_Putin follows any YouTube videos posted to reddit and responds to them with the first, most popular comment on said video. /u/Hearing_Aid-Bot responds to a redditor's use of "what" by repeating the previous comment in all capital letters. /u/PleaseRespectTables responds to threads in which a redditor uses Unicode symbols to visually "flip" a table indicating displeasure: (╯°□°)╯︵ ┻━┻. The bot will respond with a version of the table righted: ┬─┬ノ(゜-゜ノ), suggesting that it was not pleased at having the table flipped. Likewise, /u/CreepySmile-Bot posts a creepy version of a smile "ಠ‿ಠ" if someone posts the Unicode for a disapproving smile "ಠ_ಠ" as a response in a given thread. To further complicate matters, both /u/DisapprovalBot and /u/CreepierSmileBot often respond to /u/CreepySmileBot: the former with the disapproving smile mentioned above, and the latter with its own Unicode version of a creepy smile: (͡° ͜ʖ ͡°).

Some bots are programmed to respond to common phrases used on reddit as a form of play. For example, /u/Sandstorm_Bot will comment if a redditor writes, "what song is this," "name of this song," "dududu," "sandstorm," and "darude" (among others) with a link to a video for Darude's song "Sandstorm," a high-energy techno track released in 1999 (Sandstorm_Bot, 2014). /u/CIRCLEJERK_BOT responds to certain keywords (such as history, picture, girlfriend, this) with the kind of stereotypical response you might find on reddit. For example, if a user discusses Ron Paul, the bot responds with a selection of phrases, such as: "'I would vote for Ron Paul if I was still alive'—Carl Sagan" or "'Ron Paul 2012'—Joseph Kony" (CIRCLEJERK_BOT, 2012). Perhaps unsurprisingly, /u/CIRCLEJERK_BOT was banned from a number of popular subreddits in June 2012 and has not posted to the site since. Moderators have taken different approaches to allowing bots within their subreddits—many are banned, as in the case of CIRCLEJERK_BOT, to prevent discussions from derailing in the more serious-minded subreddits.[1]

As redditors are wont to push the boundaries in most ways, some accounts pose as automated bots but are actually authored by an individual. For example, "CationBot," a play on the above-mentioned "CaptionBot," posts actual advice to the AdviceAnimals subreddit in the guise of merely transcribing the text pictured in a given image macro. In a recent "Confession Bear" (Kim, 2013) posting, a user confessed to stockpiling coupons for the pizza delivery service he/she worked for so that he/she could pocket the change. "CationBot" posted a highly upvoted response, suggesting that he/she had directly transcribed the image macro, which read in part:

Actual Advice Mallard

MANAGEMENT WILL CATCH ON ONCE THEY SEE THAT EVERY CASH TRANSACTION YOU HAVE HANDLED HAS USED A COUPON …

These captions aren't guaranteed to be correct. (CationBot, 2013)

In this case, CationBot is engaging in a "game" of mimicry (Caillois, 2001) at two levels. First, CationBot mimics other bots that populate more broadly, playfully suggesting that its commentary is the result of an automatic script rather than an actual person. Second, CationBot mimics CaptionBot's posting style specifically, both by using all capital letters for the supposed transcription and by including a statement about it not necessarily being an accurate one.[2] Noticing and appreciating the multiple levels of play that CationBot demonstrates require a deep understanding of both reddit culture and the realities of the reddit platform, which allows for these sorts of scripts to populate the site.

Like their human cousin the novelty account, bots elicit a variety of responses from redditors. Some clearly enjoy their presence on the site (as evidenced by the numbers of bots that have received Reddit Gold, such as the aforementioned BeHappyBot), while others argue that they further clutter an already chaotic community. And there is a primary difference between the two—bots offer automated responses limited by their scripting, whereas novelty accounts require a redditor on the other end to participate willingly in the thread. Likewise, moderators view bots both as useful tools, often employing them for low-level moderation tasks as I mentioned earlier, and as potential nuisances. In a March 2014 posting to the AskReddit subreddit, someone asked what bots redditors loved the most. The thread was filled with misspelled Gandhis, flipped tables, and other attempts to summon certain bots. When asked why most bots were banned from many subreddits, one of the /r/AskReddit moderators responded that he/she actually liked many of them, but because of the number of submissions the subreddit received, "the saturation of bots being summoned all over the place would be extremely distracting against actual discussion." Another moderator agreed and provided more explanation: "If this submission were folks discussing the merits of each bot, why people think it's their favorite bot and what the bot does to benefit the community it'd be cool. It's not though, it's just thread after thread of bot spam" (Hp9rhr, 2014). While this may increase the carnivalesque nature of the site, it also can derail interactions, particularly in the larger default subreddits.

Novelty subreddits

Another unique feature of reddit is also the direct result of the platform's technological design. Since anyone can create a public or private subreddit for any purpose with little effort, the community often creates niche subreddits around a theme or as a result of a meme mentioned in a thread. Examples of these include /r/birdswitharms (pictures of birds edited to include human arms), /r/harlemshake (Harlem Shake videos), and /r/TheStopGirl (dedicated to posting the same animated GIF response of a young woman at a sporting event mouthing "stop" to the camera over and over). Other novelty subreddits grow out of individual comments in a thread, such as someone wondering what cats would post to reddit if they could, which resulted in the formation of /r/catReddit. A typical posting features a picture of a cat sitting in a chair at the front desk of a veterinarian's waiting room with the caption, "I've come-owdeered the awful 'vet' torture dungeon!!" For the most part, these kinds of one-off, meme-based subreddits are abandoned quickly as the collective moves on to other things. Still others live on, like the popular /r/MURICA—a satirical subreddit celebrating American exceptionalism. Redditors are encouraged to refer to Communists as "damn commies" and otherwise "Love 'MURICA" as the sidebar notes.

Some novelty subreddits grow out of reddit's role as a participatory culture platform dependent on user-generated content. /r/photoshopbattles, for example, encourages redditors to make clever or humorous adjustments to pictures uploaded by other community members. Some of these are particularly representative of reddit's collective ethos, which tends to trade simultaneously in high- and lowbrow humor with a sometimes-bizarre twist. /r/onetruegod is devoted to the actor Nicolas Cage, with images, GIFs, and Cage-related memes all finding a welcome home. Another subreddit, /r/fuckingphilosophy, uses colloquial language to ask (and answer) substantial philosophical questions. Posts include musings on Ayn Rand ("Ayn Rand what the fuck you say?!?!?!?!" MacTheMan, 2013), a description of the life of Diogenes ("Dawg, Diogenes was punk as fuck," plexico_mcbean, 2013), and the nature of epistemology ("Today this chick said some shit like 'There are some things that are known which can't be learned.'—What the fuck kind of Epistemology is this shit?" creep_creepette, 2013). Of course, other serious philosophy subreddits exist on the site, but /r/fuckingphilosophy inverts the perception that the discipline is stodgy and humorless, and certainly embodies the kind of humor popular on reddit: simultaneously highbrow and prurient.

Other novelty subreddits are parodies of popular, often default subreddits. This includes the /r/shittyask[blank] subreddits, such as /r/shittyaskscience, which offers a humorous re-envisioning of the heavily moderated and well-respected /r/askscience. /r/shittyaskscience offers ridiculous, nonsensical answers to equally ridiculous, science-related questions that other redditors pose. For example, one redditor asked, "If 200,000 people die every year from drowning and 200,000 people have already drowned this year, does that mean I can breathe under water?" (thespamich, 2013). Responses varied, with the top-voted response offering a theory involving the Roman water god Neptune and his "Ideal Sacrificial Number": "Technically yes, once Neptune has received his Ideal Sacrificial Number (ISN) he will accept no more water deaths for that year. In this case it is 200,000. The problem is Neptune is kind of a dick and he likes to change his ISN without telling anyone causing more than 200,000 drowning victims in a year …" Other subreddits within the "shitty" community network include /r/shittysocialscience, /r/shittyprogramming, /r/ShittyPoetry, and /r/shittyfoodporn (dedicated to terrible photos of good food or good photos of terrible food). Likewise, /r/shittyama is a parody of the popular /r/IAMA (I am a [blank]. Ask me anything.) subreddit, only it encourages individuals to submit the most banal aspects of their life for questioning. For example, "I just named my wifi 'Free WiFi, the code is 1234.' The code is not 1234. AMA" or "I just made a sandwich, AMA." Other postings parody elements of the typical /r/IAMA exchange. For example, a top-rated /r/shittyama titled, "I am busy all day today. I will not be at the computer. AMA" is a play on the idea that many participants in traditional AMAs do not answer questions quickly enough to reddit's collective satisfaction or tend to commit to AMAs when they do not actually have the proper time to complete them.

Yet another parody subreddit is based on the default /r/explainlikeimfive (ELI5), which is dedicated to offering simple explanations to complex questions—ones that might otherwise require a fair amount of expertise to understand. Examples of top-rated ELI5 questions include "ELI5: Why, even though I take excellent care of my current laptop, and ones in the past, does it get slower and slower as time goes on almost to the point of not being functional, and what can I do to fix it?" and "ELI5: What exactly is Obamacare and what did it change?" /r/explainlikeIAmA remixes elements of ELI5 and IAMA exchanges to create unique, playful questions and answers. For example, a redditor might request that others explain hipsters to them: "… like I'm a hipster and you're a hipster and neither of us believe or are willing to believe we're both hipsters." Or explain quantum field theory with small words: "Explain the basics of quantum field theory like I am completely terrified by words featuring any more than 4 letters." Or set up an elaborate story featuring two ghosts: "Explain why flipping the light switch constantly is scaring the home owners like I'm a ghost who died in 1776 that's fascinated with technology and you're a ghost who died two years ago that's taking out their frustration on not being able to figure out the home owner's Netflix password on me." Some threads are more pointed ("Explain to me why you just bought reddit gold for Bill Gates, like I am a poor African child about to die from malaria") and some are just heart wrenching ("Explain what is about to happen like I am your dog of 13 years who you are about to have put down"), but most /r/explainlikeIAmA submissions are offbeat non sequiturs encouraging humorous and creative responses.

Still other novelty subreddits are conundrums. There is /r/Pyongyang, which according to its sidebar "features information curated by the Committee for Cultural Relations with Foreign Countries of the Democratic People's Republic of Korea." It is a closed subreddit, meaning there is only one official submitter who also serves as the subreddit's moderator (/u/KimMyungKi). Individuals are regularly "banned" from commenting on submissions unless their words extol the virtues of North Korea and its policies. This is also true if North Korea is mentioned across reddit—redditors regularly receive automated messages telling them that they have been banned from posting to /r/Pyongyang if they question leader Kim Jong Un or the country's policies in other subreddits. And redditors who do post to /r/Pyongyang do so convincingly, extolling the Great Leader and denouncing American hegemony and its imperialist aggression against the DPRK. Small adjustments to the subreddit's style sheet further extend the idea that it is run by a totalitarian regime. No downvotes are allowed and the actual number of comments listed under each submission is fuzzed, so one posting that has only five comments, for example, will show up as having 51. When I first came across references to /r/Pyongyang early on in my fieldwork, I had a hard time understanding whether it was a parody or was actually run by a North Korean sympathizer (stranger things have happened on the internet). Reddit is an overwhelming place for new redditors, and as sarcasm and parody do not always translate well (given the strange machinations of the DPRK propaganda machine), I momentarily considered it a possibility that it was serious. Instead, the subreddit stands as a brilliant piece of satire—but is based on actual material published by the DPRK.[3]

Another subreddit that causes some head-scratching is /r/fifthworldproblems, a cousin of the relatively straightforward /r/firstworldproblems. /r/firstworldproblems is a popular subreddit where individuals post minor, everyday annoyances—things that would be considered "problems" only in developed countries. Examples might be how long it takes to turn on your HD TV or how bitterly cold air-conditioning makes your office in the summer months.

The subreddit's sidebar suggests that submissions should be "… a problem you can only have if you have money [and then] we'll feel bad for you. Then we'll feel guilty for having enough money to have the same problem." Likewise, there are subreddits for the Second World (/r/SecondWorldProblems) detailing problems one might have in the former Soviet bloc, Third World (/r/thirdworldproblems), parodying issues that arise in developing, "Third World" countries, and Fourth World (/r/fourthworldproblems), where redditors can submit the supposed problems that an undiscovered tribal culture might have. /r/fifthworldproblems suggests a Fifth World—a world beyond all that exists currently. As one blogger noted, "The fifth world is all about the everyday inconveniences of transcendence. It's peeling the veil of logic and rules and being puzzled by the surreality below. It's opening your door and finding yourself in a painting of Magritte, or worse" (devicerandom, 2012). For example, a posting might ask for advice from other redditors as they have become death and Google+ refuses to acknowledge it, or how they lost four months in twenty pounds. Subreddits of this ilk continue onward, until /r/infiniteworldproblems (which features an eye-meltingly hard-to-read CSS background).

A moderator for the /r/sixthworldproblems subreddit described the distinctions between each in a 2011 post after someone complained that the single-letter postings (for example, "EEDDDDDWWWWWWW") that characterized the subreddit were nonsensical:

> There's method behind the madness.
>
> Firstworldproblems—the woes of the western entitled
>
> Secondworldproblems—the woes of the eastern poor
>
> Thirdworldproblems—the woes of the developing untouchables
>
> Fourthworldproblems—the woes of the unrecognised and lost
>
> Each ascending number abstracts the subject's standing, and by four there's nothing left materialistically. The only thing a fourthworlder has is their perception of reality, which is subsequently abstracted in the fifth world. The sixth world abstracts the abstract, leaving you with nonsense, and the seventh world abstracts the abstraction of the abstract, bringing you to a bizzaro starting point. (happybadger, 2011)

Given the relative obscurity of these subreddits, most people would stumble on them only if they were playing the /r/random "game." Going to www.

reddit.com/r/random or clicking the "random" link on reddit's top navigation bar chooses a public subreddit at random to display, which can become both an effective way to discover less-trafficked subreddits and an addictive pastime.

RES tagging and the reddit switcharoo

Elements of play on the site are often enabled by clever modifications of the reddit platform itself. As mentioned above, for example, novelty accounts are partially a function of the large number of accounts on the site, the lengthy usernames allowed, and the low barrier of entry. This means redditors can quickly create a number of accounts on the fly with only a username and password. This encourages play around usernames and often becomes a point of discussion in threads. In addition to the often-stated "relevant username" that a novelty account or on-the-fly account relating to the content under discussion may receive, redditors will often make note of the "tag" they have given a particular user using Reddit Enhancement Suite (RES). For example, a person might say, "Haha, I have you tagged as x," or "Why do I have you tagged as x?" where "x" indicates something about previous interactions the redditor had with that person. Typically, this serves as an invitation for the tag-ee to share or link to the interaction in question (usually an anecdote of some sort). Tags can also serve as "warnings" for redditors about accounts they believe are trolls or are otherwise problematic. For example, the novelty account /u/r_spiders_link comments on postings in a seemingly normal way, but then includes a link that turns out to be an unrelated picture of a spider. Since redditors share an irrational fear ("nope") about spiders, almost inevitably one of the responses to /u/r_spiders_link will be "I am tagging you in RES so this doesn't happen again," or "I have you tagged in RES in bright red but I still clicked!" Interestingly, there is a way for redditors to include this information when they tag someone using RES, but these kinds of interactions are frequent regardless. Perhaps this has more to do with the desire of redditors to share their personal experience of reddit with others and have these experiences validated (or continue the ritual of play). However, like other forms of play on the site, it can easily create an atmosphere wherein off-topic conversations around someone's RES tag (something like, "I have you tagged as 'dog hater.' Why do you hate dogs so much?" in /r/news) dominate and derail the conversation at hand (WoozleWuzzle, 2012).

Another playful activity enabled by the reddit platform is called the "reddit switcharoo." In this long-running, collective joke, a redditor makes a joke that exchanges one object for another, and another responds with a link titled something like, "Oh, the old reddit switcharoo!" The link then links to an earlier version of the joke, and so forth. The sidebar of /r/switcharoo, which chronicles the long chain of links, describes the game this way:

> A switcharoo occurs when a witty redditor feigns ignorance about which of two subjects in a post (typically an image) is more comment-worthy.
>
> These 'roos are linked in a chain; this subreddit keeps track of the most recent link. This will help keep it in one long line, (as the creator intended) instead of branching out wildly.

One example is a recent picture someone took of his/her credit card, which he/she had customized with an image of Fry from the animated television series *Futurama* handing over a wad of cash to an unseen party and yelling, "Shut up and take my money!" In the photo of his credit card, the OP included a banana (the object most commonly used by redditors to indicate the scale of something). In response, a redditor jokingly wrote, "Big deal. You designed a banana. Looks just like every other banana I've seen. Not impressed." Another redditor replied, "Ah, the ol' reddit bananaroo," which was linked to a previous joke in the reddit switcharoo chain (hth6565, 2014). The links jump between subreddits and often there are comments about "jumping in" and "seeing how deep the rabbit hole goes." Eventually, if a redditor follows all of the links, he/she is greeted with a rage comic that discusses the designer's reason for creating the game, which was to emphasize how tired and overused the joke is. When /r/subredditoftheday asked what makes the switcharoo funny, a moderator of /r/switcharoo suggested that the joke was actually terrible and repetitive and that pointing this out was precisely its point: "A switcharoo itself is actually the **opposite** of great. It's a repetitive joke that's not funny at all when you realize how often it's told, or if you're expecting it. Hence the phrase 'Ah the **Old** Reddit Switcharoo'" (LGBTerrific, 2013). But another moderator disagreed and argued that it was actually exciting to see new redditors discovering the joke and how long the chain went:

> For me personally, it's the fact that new people continue to discover the Switcheroo every day and are fascinated by just how deep it goes. That, and people leaving messages for "future travelers". While some may consider it overplayed, it is still something that a lot of people enjoy. The fun is not in pointing out a switcharoo. The fun is going down the proverbial rabbit hole. (LGBTerrific, 2013)

The reddit switcharoo beautifully illustrates the kind of humor reddit relies on: simultaneously inventive and repetitive. It also demonstrates the kind of thinking that reddit prizes most—lateral leaps between seemingly unrelated topics—and emphasizes the community's technical prowess. While figuring out the mechanism to create a long series of links that all traverse back to the same rage comic starting point is not difficult per se, it is unlikely that many of the internet's "typical" denizens would even consider this meaningful enough to contemplate the technical requirements behind it. Following the reddit switcharoo floats you through a multitude of subreddits, mostly defaults, and time periods. I could not help but get caught up in reading other, unrelated comments on older postings when linking between stops on the switcharoo chain. The switcharoo is also the perfect joke for reddit, as it requires a certain amount of insider-y knowledge about the community even to stumble upon it, and it is just difficult enough to describe to reddit novices that it feels like an elaborate, secret handshake one might learn to gain entry into an underground club.

CSS modifications and flair

Moderators have extensive power to change the appearance (if not the functionality) of their subreddits, and this has led to another form of play on the site. For example, they can customize usernames by adding "flair" that appears only in that particular subreddit. As with RES tagging, flair can be a way for individuals to individuate themselves from other subreddit contributors—but unlike RES tagging, flair is requested by individual redditors or assigned by moderators. Like other forms of play on the site, sometimes this serves informational or purely humorous purposes and can be text or image based. For instance, /r/chicago allows users to have their neighborhood name automatically appear next to their username when they post in the subreddit. /r/AskAcademia allows redditors to include their degree information and subject area. Subscribers to /r/AdviceAnimals can choose from a selection of tiny thumbnails of popular memes, such as Business Cat or Confession Bear. Sometimes flair is a kind of reward to a user who has contributed particularly good content in a subreddit. In /r/SubredditDrama, those users who moderators judge as contributing particularly good recaps of drama have special "recap" flair.

Through modifications to a subreddit's Cascading Style Sheet (CSS) file, moderators can also create custom logos, headers, background images, and fonts and change the look of the upvote/downvote buttons for their subreddit. As mentioned in the previous chapter, some of the circlejerk meta-subreddits are the most creative in their use of CSS modifications. The moderators of /r/circlejerk and /r/Braveryjerk change their CSS files regularly—usually as a way of playfully commenting on whatever is at the top of the reddit zeitgeist at that moment. In addition, some subreddits, such as /r/AskHistorians and /r/changemyview, allow only upvotes by hiding the downvote functionality using CSS.[4] Presumably, these subreddits disallow downvotes as a way to improve the kind of conversations that appear on the site by encouraging desired comments/submissions rather than discouraging undesired ones. However, others use the downvoting mechanism as a way to make a statement about the social and political realities of discourse on reddit—a point to which I return in the next chapter.

Karma(ic) rewards

The precise mechanism of karma and the ways in which it is manipulated by reddit administrators to achieve the desired effect of privileging new, highly trafficked threads and posts on the top of the user's front page are somewhat opaque. Reddit administrators have noted that preventing spam and vote manipulation is of great importance to maintaining the site's quality, and so votes for the most popular links are "fuzzed" to prevent this from happening (jedberg, 2010). Often, particularly popular new posts do not display their karma total in an attempt to reduce the possibility of a bandwagon effect, whereby they receive more karma simply because they are already popular. And, interestingly, the reddit algorithm works logarithmically, so that the first 10 votes on a link or comment are weighted the same as the next 100 and the next 1000 and so forth (Salihefendic, 2010). This privileges and gives tremendous power to the votes of those redditors willing to browse the /new queues for each subreddit and vote on the newest material that others are submitting. Karma (also referred to as "karma points" or simply "points") is assigned based on a simple formula: upvotes minus downvotes. So a posting with 800 points might have 5800 upvotes and 5000 downvotes or 1800 upvotes and 1000 downvotes—in other words, the number of points a given posting has gives no indication of how many upvotes or downvotes it actually received—at least on the front page of the site or subreddit where this information is not displayed.[5]

Karma is a desirable reward for participating on reddit, at least if the site's discourse is to be believed. Subreddits such as /r/FreeKarma exist solely to upvote a redditor's submissions, for those whose accounts are new or those who just want the karma.⁶ A number of web sites (such as karmawhores.net) calculate the accounts that have the most karma, offering insight into who might be loosely classified as "reddit celebrities." These users are often mentioned in reddit threads (which, with Reddit Gold, sends a message to said user's inbox), which often results in their appearance—and a bounty of karma for whatever contribution they make.

And yet karma is equally derided as a useless metric or at least one unworthy of all of the attention that redditors pay to it. Subreddits such as /r/KarmaConspiracy highlight postings that redditors believe may have been fabricated solely for the purpose of gaining karma. For example, after the popularity of the safecracking incident mentioned in the last chapter, numerous other safes started appearing (and being upvoted) to /r/pics. These would then be "opened" by the OP, resulting in an update posting with information about its contents that would garner even more upvotes. /r/KarmaConspiracy would link to postings like these and "research" the factual basis for the OP's claims.

My informants suggested that redditors are keenly aware of what is popular with other redditors and often take this into account when posting content to the site. As Armani noted,

> Redditors know what the common stereotypical redditor likes, so they'll intentionally write comments that will appeal to the masses, eventually leading to garnering upvotes (e.g. hehehe lizard, puns, making a forever alone reference). Similarly, people know what kind of submissions that the typical redditor would enjoy as well. For example, even if an experienced redditor truly found Carrot Top funny, they probably wouldn't post a Carrot Top joke in /r/funny because the overwhelming majority of redditors can't stand Carrot Top and will downvote the submission.

Comparing karma points is often discussed in threads. Redditors will often note when a comment has been their highest upvoted one or will discuss each other's karma points, saying something like, "2,000 link karma and 3,000 comment karma for an account that's six months old—not bad!"

This creates a kind of double bind. Redditors desire their contributions to the community to be recognized through upvotes, but obtaining this is most easily done by invoking certain already-popular memes—a direct contradiction of the site's emphasis on original and unique content. And despite differences in opinion about the role that karma plays in an individual's experience of reddit and the kind of posts that are rewarded upvotes by the community, there seems to be real power in these "invisible internet points." While individuals assign multiple and sometimes contradictory meanings to karma, it is clear that it *does mean something*, beyond simply an algorithmic way of demonstrating what's most popular on reddit. Karma offers a concrete articulation of the perceived value of contributions that a particular redditor makes to the community, which echoes Alex Halavais's (2009) findings in his analysis of voting and contribution patterns on the social news-sharing site Digg. However, my interviewees had mixed feelings about the value of karma. On the one hand, some found it a useful way to quickly assess the quality of posts to a subreddit or comments in a thread. As Jessie argued,

> It's kind of like if your friend tells you about someone you have never met. If your friend says something either positive or negative about this person, when you actually meet them you will have your friend's thoughts floating around in your head. If a comment has a lot of downvotes I will automatically think poorly of it, even before I read it. Reverse with the upvotes. I then read the comment and make my own judgment.

But on the other hand, as Marley suggested, karma points were mostly useful only as a way to organize content on the site:

> Well, karma in comment sections doesn't really accomplish anything besides ordering the way in which comments are presented on the page. Even then, you have the option of reordering how comments are sorted. People sometimes claim that downvoting sensors [*sic*] unpopular opinions, but really it just changes how they are prioritized on the page in the default sorting option; basically just meaning that it takes a little more work to find them.

However, the ease of upvoting and downvoting material on reddit means that many posts with a large amount of karma are perceived as superficial or pandering to the reddit "hivemind" mentioned in the last chapter. One informant, Finley, commented,

> The positive thing about karma is that content that receive a lot of karma by upvotes are often interesting. However they are mostly interesting in a superficial way ... My most recent high earner was a humorous fan fiction story about a famous soccer player could break out of jail and join a different team. However what value did it add? ... I would guess that most posts that did improve my quality of life were found [in] a sub reddit instead of the front page. Those naturally do not have as much karma.

Finley was most intent on using the community to learn new things—as a way to gain knowledge rather than simply pass the time. Thus, for this individual, karma often indicated superficially interesting content but did not meet his need to acquire new information.

Other informants suggested that karma is important to community members, even if "publically disavowed." As Cameron noted, it is a point of pride to have your postings upvoted: "… when I posted my art to /r/StarWars, I got some upvotes and some downvotes. I wish whoever downvoted would have told me why, but the upvotes let me know that people liked what I did and it made me feel good." This sentiment is echoed by many other redditors in public threads—suggesting that gaining karma points is pleasurable and serves as some sort of indicator of acceptance or agreement by the larger reddit community. If further proof is needed regarding the importance of karma, the rise and fall of reddit celebrity /u/Unidan is one example. In July 2014, his primary account was shadowbanned for vote manipulation. He admitted to creating several alternate accounts that he would use to log in and upvote material he had posted and downvote that with which he disagreed. (The entire drama was recounted in excruciating detail by, of course, SRD subscribers [the_real_ stabulous, 2014]). This demonstrates that karma can be intoxicating, even (or especially) for those accounts that have the most of it.

Karmawhoring and shitposting

"Karmawhoring," or what the community decides is merely an attempt to gain karma, is derided by redditors, perhaps because it is viewed as unsportsmanlike. What precisely constitutes "karmawhoring" is contested, but it seems to be related to posting content that panders to reddit's interests. Pets, particularly cats, are referred to as "karma machines" and are perceived as the primary mechanism by which one might obtain karma. While pet pictures frequently fill the top space on the front page of both /r/aww and /r/funny, comments on such posts often suggest that these sorts of posts are not worthy of the attention they receive, as they are often staged or simply "guaranteed" to gain a lot of upvotes regardless of their quality. Likewise, reddit has mixed feelings about the "cake day" posting—a post that's usually of a pet or cute female or nostalgic pop culture item—submitted on the redditor's yearly anniversary.[7] "Cake day" posts to the default subreddits are typically rewarded with lots of upvotes, and thus the OP receives lots of karma points. This occurs despite the fact that many comments about the post usually suggest that the OP should not receive such riches and may even be karmawhoring, just because it is his "cake day."

OPs are also accused of karmawhoring when submitting content that is likely to be faked or otherwise manipulated. For example, individuals often submit pictures of amusing notes they receive from others in /r/pics—ones that might be from a neighbor thanking them for some good deed—that could easily be faked. And such accusations are made semiregularly, with the assumption that achieving a kind of reddit notoriety is enough of a goal for folks to fake something as relatively inconsequential as a kind note they received from a neighbor.

As mentioned earlier, image-based content is regularly upvoted and recycled. Looking at /r/shitpost (a repository of popular postings that some redditors consider especially questionable in terms of their merit), almost all of them are images. Unlike "karmawhoring," in which the original poster (OP) is called out for posting something solely because it is perceived to appeal to reddit's sensibilities, suggesting that a submission is a "shitpost" is an insult directed at both the poster and those who upvote it. Shitposts are almost invariably low effort and pandering: an example is posting a picture of the cover of a newly purchased *World of Warcraft* game with the title, "Just bought this today, will be my first time playing" to the *World of Warcraft*–dedicated subreddit (/r/wow). This posting received 1090 points (domo-loves-yoshi, 2014). The term "shitpost" is likely borrowed from 4chan, where it is described as "knowingly contributing low quality, off-topic, or ill intentioned posts" ("4chan Frequently Asked Questions," 2014). Usually, someone in a given thread will simply make the comment "shitpost" or "/r/shitpost" to suggest that the posting has or should be submitted to /r/shitpost—marking it as a crappy posting unworthy of the attention it is receiving from the reddit community.

Importantly, accusations of both shitposting and karmawhoring suggest karma points are meaningful in and of themselves. If karma were entirely meaningless, accusations of both would be pointless and would be infrequent (if mentioned at all). I suspect that it is not merely about the visibility that numerous upvotes of a given posting or comment bring, but that number itself has intrinsic value. Redditors view karma as an indicator of how much an individual has contributed to the community as a whole while also becoming suspicious of those who continually reap karmic rewards. And a non-celebrity redditor who hits the karma jackpot within a few days of creating an account is also viewed with some suspicion, perhaps because that person is perceived as not being a true "member" of the community. I should also note that it seems as though accusations of karmawhoring and shitposting grow in the summer months (which I suspect is part of the "school's out for summer" phenomenon discussed in the last chapter); whether or not this is actually true is debatable.

Reputation and voting behavior

It is important to note how redditors' behavior and site engagement are likely shaped by the site's pseudoanonymity, the mechanics of karma, and tools such as Reddit Enhancement Suite (RES). As one /r/circlebroke poster notes, RES's tagging functionality and its ability to show the number of upvotes or downvotes assigned to another reddit account likely influence future voting behavior:

> The way RES tallies votes is directly influencing the way that I vote in a thread. Not only does the bright green draw my attention to people that I have upvoted in the past, but I no longer find myself voting on the merits of the comment. When I see something blatantly wrong or unhelpful I found myself *hesitating and reluctant* to downvote if they have a green tally next to their name.
>
> Not only that but I find that I've upvoted comments from those people that provided very little to the conversation. I shouldn't have to explain that the reverse is true for those who are now in the negatives. (BronzeLeague, 2013, emphasis in original)

BronzeLeague's experience mirrors my own. It was not until the end of my project that I started tagging people in earnest through RES or even really noting the points I had assigned them in the past. This is likely because I often accessed the site through BaconReader (an Android application for accessing reddit) on my phone, which does not include RES functionality, so I had very little sense that my prior voting behavior overtly shaped later votes. But I was aware of certain usernames that I saw repeatedly on the site, especially those that seemed to appear regularly on the site's default subreddits. These were not just reddit celebrities like /u/Unidan, either, but everyday redditors whose comments and postings I would greet with amusement or derision and upvote or downvote accordingly. Regardless of the method of accessing reddit, it is likely that longtime participants become aware of regular posters they see—especially those within the smaller subreddits they visit regularly. So despite extensive encouragement by reddit moderators and administrators for users to let reddiquette influence voting behavior and guide their "play" on the site, there is little evidence that it actually does.

BronzeLeague's posting also contained another revelation about the interrelatedness of the reason that easy karma-grabbing material such as reaction GIFs is popular on the site despite individuals decrying the fact that nobody really cares about the "invisible internet points" that karma represents—both for those receiving it and for those doling it out:

> I used to find it so hard to believe that people care about karma. This whole experience has demonstrated to me that like it or not, I do. You do. You can't help but care. I'm strongly considering uninstalling RES so I can go back to feeling self-righteous about how Karma doesn't matter to me and dispensing my votes without having to guess whether I like what was said or who said it.
>
> Reddit is the best possible example of what keeping score in discussions leads to. (BronzeLeague, 2013)

For most redditors, RES serves as a useful suite of tools that, like its name, "enhances" their reddit experience. But such tools, like all technologies, are not value neutral.

Voting behavior also ties more broadly to site participation. The idea that there are many individuals who either choose not to vote on a posting or cannot vote on reddit echoes the relative paucity of participation on these kinds of platforms more broadly. As Dakota notes,

> …the 90-9-1 rule is interesting here. In theory, for every up- or downvote I've received, ten people have seen my post. While that's probably not true across the board on reddit (because very little is), I think it's interesting that there's this huge "silent" audience out there on the internet. It's simultaneously powerful, disappointing, and disconcerting that so many people choose not to interact—and that so few people choose to contribute, yet still manage to generate massive amounts of content every day.

For this individual, participation within the community is critical to its survival, and yet as he noted, it is humbling to see how much content is produced despite such a low rate of participation on the site as a whole.

Reddit rules

There are rules governing how play occurs on reddit. Rules set the boundaries in gaming and play spaces, providing structure and challenge for players. As Jesper Juul (2005) argues, "We may associate rules with being barred from doing something we really want, but in games, we voluntarily submit to rules. Game rules are designed to be easy to learn, to work without requiring any ingenuity from the players, but they also provide challenges that *require* ingenuity to overcome" (p. 55). However, in all games, there is a difference between the articulated rules of play and the actual rules that players use to guide their behavior (Salen & Zimmerman, 2003). The rules shaping behavior on reddit are no different.

There are only five enforced rules that will cause site administrators to ban someone's account.[8] The first restricts users from spamming the site. Spamming in this case can be either linking to one's own content in an egregious manner or repeatedly submitting the same content over and over again. The second restricts vote manipulation. In the rules section, reddit administrators note that while sharing links with friends is fine, asking for upvotes for whatever reason is not. The third rule restricts the sharing of any information about private individuals' "real life," links to their social networking presence, or private information about public figures (according to this rule, a link to a senator's public home page is fine; sharing her personal cell phone number would not be)—all of which would constitute doxxing. The fourth rule prohibits the posting of child pornography or sexually suggestive material containing minors. The final rule restricts individuals from breaking the site or otherwise interfering with its operation when making calls using the reddit API, which is used when creating a reddit bot, for example ("Rules of reddit," 2014).

At the very bottom of the list of reddit rules is a small addendum: "You should also be mindful of reddiquette, an **informal** expression of reddit's community values as written by the community itself. Please abide by it the best you can." As discussed in the last chapter, reddiquette provides suggestions about what content is encouraged and what content is not allowed on the site ("reddiquette—reddit.com," 2013). Despite the explicitness of these rules, many of my respondents noted that they thought no one else on the site, besides themselves, actually followed them. And some openly admitted that they, themselves, did not follow reddiquette when it came to their voting behavior. Jessie, for example, is "not a fan of dogs," pornographic material, and the Confession Bear meme and actively downvotes this kind of content when it appears on the front page.

Besides the explicit rules that reddiquette offers, which individuals may or may not choose to follow, and the more implicit informal policing that occurs as a consequence of the site's mechanics, individual subreddits each have their own set of rules enforced by moderators. This can include specific content guidelines for submitted links (for example, /r/Games does not allow memes or GIFs) or standards for the discussions/interactions that occur in threads, (/r/chicago has a stated policy against racism, or the spoiler policy enforced in many movie or television show subreddits), or finer-grained rules around that particular subreddit's subject matter, which may be a direct result of "spinning off" from another subreddit. For example, /r/fffffffuuuuuuuuuuuu (f7u12) is dedicated to publishing rage comics (opspe, 2013). /r/f7u12 has relatively loose rules around what material can be submitted. Therefore, content in the f7u12 subreddit has moved away from the "classic" four-panel model to feature other, longer comics. Because some redditors felt that f7u12 was too diluted with non-canon rage comics, /r/classicrage was created, with rules that prohibit any comics that are more than four panels or do not end with the appropriate rageguy face. These rules and how well they are enforced by moderators and followed by subscribers shape the content, interaction, and tone of individual subreddits. It is what makes the experience of /r/Games (dedicated to videogames with strict content-based guidelines) substantially different from /r/gaming (a more popular, default subreddit dedicated to videogames with much more relaxed moderation).

Inverting play

It is not surprising, then, that several different kinds of inversions of play can be found on reddit that directly (or indirectly) challenge the rules of the playspace. Multiple game studies scholars have discussed the importance of inverting rules and creating new, often subversive ways of playing (Aarseth, 2007; Consalvo, 2007; Postigo, 2010). Alexander Galloway's (2006) concept of "countergaming" is a useful way of understanding how playing with a game's mechanics or visual elements can disrupt and refocus our attention as players. Countergaming is "oppositional cultural production" in which the ludic and narrative spaces of games are transformed, often for artistic, political, or simply playful reasons (Galloway, 2006, p. 109). On reddit, one way to invert the rules is not to play or participate at all. Many individuals visit the site on a daily basis but do not have an account or simply choose not to vote. As with other participatory culture platforms, there is a sense that a much greater number of individuals on reddit are participating implicitly rather than explicitly (Schäfer, 2011).

While play on reddit is governed by both reddiquette and individual subreddit guidelines, there are times when even strict rules governing content are temporarily relaxed. Two examples illustrate this point. Every month or so in /r/gaming, a default subreddit that allows only content related to videogames, a link will be submitted and upvoted to the top of the subreddit that reads something like, "The mods are asleep! Quick, post board games!" (TH4N, 2013). Inevitably, a long thread will follow, with people sharing favorite games from their childhood or extolling the virtues of current board game favorites. Participants clearly revel in the kind of subversive play that threads like these often exude, even if they are directly against the stated rules of a given subreddit. That these sorts of threads happen sporadically but regularly in /r/gaming and yet are not deleted automatically by moderators suggests that much like other games, the rules are subject to a fair amount of interpretive flexibility. And these moments underscore the nature of reddit as a *sanctioned* carnival, where both reddit members and moderators are in on the joke.

Another example of the ways in which even strict content rules are broken in the name of play occurred recently on /r/apple (dedicated to discussions about Apple products). A posting (since deleted) posed a question worded something like, "What do I do with the leftover almond meal after making almond milk?" (hoontur, 2013). Later, the original poster (OP) edited his original post to clarify that he swore he was in a different food-related subreddit and apologized for cluttering /r/apple with such an off-topic question. Normally, this kind of posting would have been downvoted (at the very least) or deleted by the subreddit's moderators. In this case, perhaps because the question was such a departure from the usual content posted to /r/apple, commenters posted likewise off-topic responses ranging from long treatises on the Third Estate of France to arguments about whether or not Greedo shot Han Solo first in *Star Wars* to statements about Boeing 747's popularity for international long-haul flights to a well-known quote from the television show *Seinfeld*: "These pretzels are making me thirsty." Others responded jokingly with Apple-related advice about resetting the iMac G5's SMU (System Management Unit) or how one might use the almond meal to create a paste using milk and toilet paper to jam into a MacBook Pro so that the fan might run more quietly.

Both of these examples highlight an important aspect of the reddit experience and a familiar one to scholars who research the community nature of play in gaming spaces (see, for example, Nardi, 2010; Pearce, 2011; Taylor, 2009): play is a collective activity involving multiple people—or, at least, multiple accounts. While an individual actor may have instigated these moments of play, gaining traction within the community requires collective action to support the playful activity. And these examples highlight the reddit community's carnivalesque nature and love of the non sequitur. Inversions of reddit's play can also be invoked for political reasons, a point I discuss in the next chapter.

Trolling

From the outside, it might be easy to mistake these examples of inverting play for sophisticated trolling. However, there seem to be distinct differences between the two. Many trolling comments on reddit exude insincerity and derision. Additionally, trolling behavior is often repetitive, insistent, and, most important, *identified* as trolling by other community members (Bergstrom, 2011). But in the case of the threads mentioned above, the originating comment suggests sincerity, and in the case of the board gaming example, playfulness. Collective enjoyment, rather than individual self-promotion, seems to be a defining characteristic that distinguishes this kind of play from mere trolling.

Another way to invert the rules of reddit is to troll in an attempt to gain negative karma points through downvotes. Trolling behavior includes submitting content and/or comments that are perceived by the community as purposefully inflammatory, off topic, or are otherwise intended to elicit a negative response (see Bergstrom, 2011) for one detailed example) simply for the LULZ (fun or amusement) (Phillips, 2015). Some of this behavior is likely the result of spillover from communities such as 4chan's /b board, where trolling behavior is commonplace (Bernstein et al., 2011). The problem with trolling is that it is self-reinforcing, as many attempts to troll are rewarded with downvotes, which is, presumably, just what the person doing the trolling desires (hence the common rejoinder on the site, "don't feed the troll"). Until it was banned in 2012 (Morris, 2012b), an entire subreddit devoted to trolling reddit existed—/r/gameoftrolls (which took its name from the popular fantasy series *Game of Thrones*)—solely to antagonize and hijack threads in other subbreddits in an attempt to accumulate negative karma. It was set up as a kind of game, where subscribers would attempt to troll other subreddits without being identified as a /r/gameoftrolls player (definitelymyrealname, 2013).

Trolls are easily tagged through RES if identified, which might allow people "not to feed the troll" downvotes inadvertently. That being said, it is difficult not to downvote or respond to trolling comments (as trolling is most visible in comment threads) as they are intended to elicit anger or disgust in those reading them. As Whitney Phillips argues, the "mask" of trolling is a repudiation of many of the logics of social media more broadly, namely, the notions of "transparency, connectedness and sentimentality" (Phillips, 2015). These values underscore much of what makes an online community like reddit functional and enjoyable for participants, which is one reason /r/gameoftrolls was banned (vote brigading/manipulation being another). While the playful rejection of rules or inversions of play on reddit may flirt with these modes of being, it often lacks the persistence and the "cool rationality" that marks trolling behavior (Phillips, 2015). However, inversions of play on reddit are less about the LULZ and more about pushing and determining the boundaries of the playspace.

What reddit can teach us about play

All platforms shape and constrain the kinds of interactions that occur on them. What makes reddit unique is not the fact that novelty accounts exist (as they also do on Twitter) or that memes are regular parts of site discourse (as they are often shared on social networks such as Facebook), but that these playful interactions move beyond "mere play" to reflect the cultural values of the community. In reddit's case, this means that cleverness and creativity, as in other spaces of geek culture, are prized above all else, along with plenty of crass and base humor. And, despite having a vast number of subreddits with their own subcultures, rules, and norms, reddit is united in playfulness—even if that play is subversive or objectionable.

Redditors' intense and often serious interest in play mirrors their geek pedigree. And yet the almost religious devotion with which some in the community treat their carnival is both awe inspiring and perplexing. Is this how they perceive and approach other aspects of their lives? After all, this is a place where after months of fieldwork, I find myself questioning the intent of a subreddit such as /r/outside (dedicated to sharing images and stories that treat the "real world" as a game)—Is it a parody? Is it serious? Are there people on reddit who actually believe they can divine some magic Konami code that unlocks fame, fortune, and relationship success?

But it is not simply about reddit's predilection toward geekery that makes play possible. Much of reddit's play is critically dependent on the platform's support of pseudoanonymity, its open code base, and the willing, free labor of participants. Both novelty accounts and bots could not exist in a space that was entirely anonymous, nor in one that required a "real name" for participation. In addition, reddit's open API makes this kind of play possible—and the fact that play is a valorized way by which many redditors interact with one another makes it desirable.

However, there are issues with viewing reddit solely through a lens of play. It is a critical language of reddit, but it coexists equally with the language of cynicism and drama mentioned in the previous chapters. But to dismiss reddit as "mere" play is to give both the individuals involved and the concept short shrift. It also is not an uncontested concept, as my discussions in the next chapter of the ways in which certain communities on reddit have mobilized around the concept of play to make a political critique of the platform demonstrate.

Notes

1. For the curious, several semi-comprehensive lists of bots on reddit have been posted—one is located at http://www.reddit.com/r/BotParty/wiki/index. There is also a subreddit dedicated to introducing and discussing bots: /r/botwatch. Most meta of all is a bot someone wrote to find and describe other bots (Plague_Bot, 2014).
2. To make matters even more complex, CaptionBot later revealed that he/she was also impersonating a bot by transcribing memes by hand rather than through the automated script that the account's username and behavior would indicate (CaptionBot, 2013).
3. There is also a serious (non-parody) subreddit about North Korea (/r/northkorea), which serves as a repository for news and information about the country.
4. However, any reddit user with Reddit Enhancement Suite (RES)—a popular browser add-on that allows one to ignore a subreddit's style if desired—can subvert these kinds of changes made to a subreddit's CSS file. This means that someone can circumvent the "no downvoting" rules of a subreddit if he/she so chooses.
5. Until June 2014, if a user had RES installed, she could view the number of upvote and downvote totals on the front page of the site and subreddits for postings. At that point, RES also displayed the number of upvotes and downvotes a given comment had received—a feature "vanilla reddit" did not. This changed in June 2014 when vote totals were removed from the site in favor of reporting a percentage or a vote count (for comments).
6. Some subreddits require accounts to have a certain amount of karma or be of a certain age before the account holder can submit things to them, presumably to avoid spam or trolling.
7. So named because of the cake icon that appears next to the redditor's username on that day indicating that it is the anniversary of the date that account joined reddit.
8. To prevent spammers and others from continually creating new accounts to get around bans, reddit administrators often "shadowban" accounts—that is, they make it possible for that account to post to reddit, but the account's submissions are never seen by anyone else on the site. If someone visits the profile (/user/) page for a shadowbanned account, she would see a message that the user no longer exists.

· 6 ·

OPEN PLATFORMS, CLOSED DISCOURSE

At a 2013 SXSWi[1] panel, "It's Reddit's Web. We Just Live in It," three journalists offered their insight on the community. After spending some time covering the community's admirable activism around SOPA and CISPA, two attempts by US legislators to restrict internet freedom, the panel moved on to other, not-so-positive aspects of reddit—namely, the small segment of the community dedicated to posting images of unsuspecting women (/r/creepshots), previous problems with child pornography, and the sexism and misogyny rampant on the site as a whole. During the Q&A session for the panel, a conference attendee commandeered the microphone to rattle off a list of positive actions the reddit community had engaged in, including the largest Secret Santa gift exchange, relief efforts in Haiti, and other charitable giving by members.[2] Rebecca Watson, founder of the blog Skepchick, stepped in to suggest that the panel had covered the good parts of reddit but that the negative aspects of the site still needed to be addressed, opining, "The good things that reddit and redditors do doesn't negate the inherent problems of reddit. And I think it's more important, especially in a site that promotes all the good it does—it's crucial for them to look at the problems in an honest and open way and try to address those." After asking whether the individual (later identified as Imgur[3] founder Alan Schaaf) agreed, he responded affirmatively if reluctantly, adding, "I also think that for a panel focused around reddit, you should have had some reddit users." His indictment of the panel's makeup was met with both applause by the audience and protests by the panelists, one of whom (Gawker's Adrian Chen) said, "But we do use reddit." This exchange brought to the fore a long-simmering tension between journalists/critics of reddit and those who would consider themselves "redditors" and the ways in which this boundary is discursively constructed by members of each of these groups quite differently. It also brings to the fore questions of "insiders" and "outsiders" within the space and who is permitted to critique what happens on reddit.

Geek culture and hegemonic masculinities

Reddit is informed by and reflective of larger dimensions of geek culture. As I have discussed in the previous chapters, everything, from the popular culture objects that serve as agents of nostalgia for the community to the kind of humor that pervades the site, demonstrates that reddit is a primary site where this kind of culture is performed. This is not to suggest that there is a solitary "geek culture" that we can point to; as with other forms of culture, what binds individuals who might self-identify as "geeks" is emergent and fluid. However, an exploration of this culture and its complicated relationship to masculinity is necessary to understand how reddit's technological logics may implicitly disenfranchise certain populations.

Scholars have used hegemonic masculinity as a sensitizing concept for understanding how issues of gender and power intersect. Early formulations in the 1980s of the term suggested that "hegemonic masculinity was understood as the pattern of practice (i.e., things done, not just a set of role expectations or an identity) that allowed men's dominance over women to continue" (Connell & Messerschmidt, 2005, p. 832). As R. W. Connell (2005) argues, there are multiple masculinities, requiring an understanding of the ways in which class, race, and sexuality (as well as age and [dis]ability) intersect and transform one another. Still, these multiple masculinities do not exist in a vacuum; they are informed by and must contend with hegemonic masculinity. Connell suggests, "'Hegemonic masculinity' is not a fixed character type, always and everywhere the same. It is, rather, the masculinity that occupies the hegemonic position in a given pattern of gender relations, a potion that is always contestable" (Connell, 2005, p. 76).

At the same time, technical expertise has long been intertwined with hegemonic masculinity. Feminist scholar Judy Wajcman argues, "… technical competence is central to the dominant cultural idea of masculinity, and its absence [is] a key feature of stereotyped femininity" (Wajcman, 1991, p. 159). The computer programmer, for example, is implicitly coded as male, whereas the technically challenged (or incompetent) computer user is coded as female. This runs contrary to the actual history of computing—early "software" was run by women (the so-called ENIAC girls of the 1940s and 1950s), who set up machines to run computations (Abbate, 2012; Ensmenger, 2010; Misa, 2010). Later, in a bid to expand the potential pool of programmers who were desperately needed in the US post–WWII boom, employers relied on personality profiles to attract new employees. Unfortunately, these created a circular problem whereby "the primary selection used by the industry selected for antisocial, mathematically inclined males, and therefore antisocial, mathematically inclined males were overrepresented in the programmer population; this in turn reinforced the popular perception that programmers *ought* to be antisocial and mathematically inclined (and therefore male), and so on" (Ensmenger, 2010, pp. 78–79, emphasis in original).

Programming and technical expertise have been socially constructed as a masculine profession and pastime. So geek masculinity therefore privileges the white, able-bodied, young straight cisgendered male over the woman of color, for example, or the homosexual older man, or the disabled trans* woman. This is not to say that these individuals are not active in geek culture but that they remain marginalized, relegated to its fringes, and frequently silenced. This becomes obvious when an individual challenges dominant assumptions about, for example, the nature of women's roles in video games (as *Feminist Frequency*'s Anita Sarkeesian did [Lewis, 2012]) or objects to the overwhelmingly white male nature of panels that make up many of the top technology conferences (Imafidon, 2013) or questions the passive acceptance of sexist talk that characterizes the audience chatter during conference presentations (as Adria Richards did [Liebelson & Raja, 2013]). It is at these moments that a vocal portion of the community will often not only suggest that these criticisms are off base but will also complain that those raising these questions are not really a part of the geek community and therefore should have no right to speak at all. More disturbing are the violent threats and slurs hurled at these individuals—especially at women and particularly women of color—suggesting that their very existence is unwelcome and somehow threatens the geek community to its core. This creates a feedback loop in which individuals may find aspects of geek culture unwelcoming and may serve to further marginalize certain voices within the community, as they are viewed somehow as interlopers and not as true "geeks." This becomes further complicated by the image of the redditor/geek as a sexually frustrated straight male (a point discussed in Chapter 3), wherein the image of the female is both desired and feared or viewed as a threat.

As Lori Kendall (2002) argues, the "nerd" or "geek" has been subsumed within contemporary notions of hegemonic masculinity. She notes,

> Since the 1980s, the previous liminal masculine identity of the nerd has been rehabilitated and partially incorporated into hegemonic masculinity. The connotations of this term thus vary on the social context. As an in-group term, it can convey affection or acceptance. Even when used pejoratively to support structures of hegemonic masculinity, it can confer grudging respect for technical expertise. (Kendall, 2002, p. 81)

Kendall goes on to suggest that "the nerd stereotype includes elements of both congruence with and rejection of hegemonic masculinity. Its connection to a reconfiguration of middle-class male masculinity partly redeems its masculine project" (p. 82).

Reddit's male gaze

To be a woman on reddit is to contend with a kind of "male gaze." In Laura Mulvey's (1989) original formation, movies often rely on a particular grammar that reinscribes straight male possession of and dominance over women. The process of watching a film becomes one of voyeuristic pleasure, whereby women are on display as the objects of male desire.[4] This is commonly seen in Hollywood films, for example, which rely on a visual and narrative grammar wherein females are relegated to being passive love interests or are presented as potential threats for the male protagonist to subdue. The prototypical example of the male gaze in action can be seen in Alfred Hitchcock's *Vertigo*, in which the protagonist (Scottie, played by Jimmy Stewart) follows, rescues, and loses a love interest (Madeline, played by Kim Novak), only to remake another woman in her image and eventually lose her. The camera becomes an extension of Scottie's vision, gazing voyeuristically and obsessively, but unseen, at Madeline. As audience members, we are complicit in this voyeurism; we are unseen as we watch Scottie watch Madeline. Moving beyond the realm of film, the concept of the male gaze has been applied to online environments to understand the ways in which women can be objectified within these spaces (see, for example, Adam, 2002; Attwood, 2011; Senft, 2008; L. L. Sullivan, 1997).[5]

We can expand on the idea of the "male gaze" through Martha Nussbaum's (2010) work on internet misogyny. She offers an insightful explication of the multiple ways in which women are subject to forms of objectification. Nussbaum argues that women and other marginalized groups are made into objects and denied agency in seven distinct ways:

1. *Instrumentality*: the objectifier treats the object as a tool of his or her purposes.

2. *Denial of Autonomy*: the objectifier treats the object as lacking in autonomy and self-determination.

3. *Inertness*: the objectifier treats the object as lacking in agency, and perhaps also in activity.

4. *Fungibility*: the objectifier treats the object as interchangeable (a) with other objects of the same time, and/or (b) with objects of other types.

5. *Violability*: the objectifier treats the object as lacking in boundary integrity, as something that it is permissible to break up, smash, break into.

6. *Ownership*: the objectifier treats the object as something that is owned by another, can be bought or sold, etc.

7. *Denial of subjectivity*: the objectifier treats the object as something whose experience and feelings (if any) need not be taken into account. (Nussbaum, 2010, pp. 69–70)

Citing the work of philosopher Rae Langton, Nussbaum suggests additional elements of "objectification" that encourage a vision of the object as mere body and appearance, and result in the silencing of the object in question. And, most important, the female being objectified in these cases is subjected to an intense amount of shame about her agency and very existence. This is common for all powerless groups, as Nussbaum suggests, but is particularly salient for women who are seen as mere objects. As she argues,

> Despite its ubiquity, objectification is a very extreme form of shaming. Shaming always involves giving someone a stigmatized, spoiled identity. But the gap between human and thing is so vast, so objectification involves an extreme form of spoiling: a statement that the person no longer counts in the universe of the human. (Nussbaum, 2010, p. 73)

In the case of reddit, a very particular male gaze dominates, one that reflects the site's role as a space in which geek masculinity is valorized. And the technological politics of the reddit platform, where highly upvoted content is most visible and contradicting views are often downvoted and thus less visible, often works to reinforce a vision of women as passive objects rather than as active agents. The reddit male gaze manifests itself in a number of ways, both obvious and subtle. For the purposes of this discussion, I consider three cases. First is the prevalence and popularity of NSFW (not-safe-for-work) content and subreddits on the site. Second are subreddits and site commentary that explicitly advocate an anti-feminist stance. Third is the portrayal of women in a number of popular image macros in /r/AdviceAnimals. While these cases do not describe the entirety of the reddit "male gaze," I believe they offer a partial glimpse of the ways in which women are objectified on the site.

NSFW subreddits and content

The most obvious demonstration of the reddit male gaze can be seen in the site's numerous not-safe-for-work (NSFW) subreddits and content. For first-time visitors who have yet to curate their feed, NSFW material is quickly apparent. Most of it falls into two categories. The first is NSFW material featuring medical imagery or stories involving bodily functions. This might be content typically featured in /r/WTF or /r/popping (a subreddit dedicated to videos and GIFs showing skin ailments such as pimples being popped)—sometimes referred to as NSFL, or "not safe for life." The second category, and my focus here, is material that is explicitly sexual in nature. While this for the most part means pictures (and mostly pictures of women), it can also be threads in /r/AskReddit or /r/sex that discuss sexuality. And it may be images that are not explicitly sexual in nature (in that they feature clothed women) but are meant to titillate male audiences. An example of this is the subreddit /r/gentlemanboners, which features fully clothed, slightly provocative images of female celebrities such as Jennifer Lawrence and Scarlett Johansson. The title of the subreddit alone offers a glimpse into the subtle ways in which the reddit male gaze works. Since only photos of women wearing evening or daytime wear (rather than swimwear or no clothes at all) are featured, the subjects are "classy" and, thus, the viewer is a "gentleman." This is further evidenced by the subreddit's description, which reads in part, "Elegant, graceful, timeless women are one click away. Be gone with you and your slutty, trashy, & whorish succubi." At the same time, the images are shared precisely because these are attractive women openly intended to provide the viewer with sexual satisfaction (hence the "boners"). The juxtaposition of the words "gentleman" next to "boners" in some ways perfectly encapsulates reddit's formulation of geek masculinity—women are objects to be admired and exist for male satisfaction, but the men doing the ogling are nice, classy gentlemen wearing, metaphorically speaking, monocles and top hats.[6] Likewise, /r/WatchItForThePlot is described as a subreddit "… for pictures, gifs, videos, etc. from movies or television shows that we watch for the, um, 'plot'." Titles of postings feature the word "plot" as a code for breasts or body as in, "19 year old Megan Fox showing us what a good plot looks like on Hope and Faith" or, "True Detective knows what a good plot truly needs (Alexandra Daddario)." Again, the emphasis here suggests that those posting and consuming these images are admiring the female form surreptitiously and thus respectfully. As explored in Chapter 3, the stereotypical redditor is viewed as sexually frustrated and, quintessentially a "nice guy." The "nice guy" often sublimates his desire for sex by offering emotional support for a woman in the guise of being her friend ("Nice Guy Syndrome," 2014). He then becomes frustrated when his romantic interest, whether stated or not, in said female friend is rebuffed; thus he is "friendzoned," or relegated solely to a platonic role in the woman's life (Nicholson, 2011). Feminists have taken issue with the "friendzone" concept, as it is passive-aggressive, suggests an ownership of or entitlement to women's bodies, and is often deployed by "nice guys" who are actually deeply misogynistic in their attitudes and/or behaviors (Lynn, 2013; Riordan, 2013; yeti-detective, 2013).

Other NSFW subreddits feature standard pornographic clips (for example, /r/NSFW_GIF, /r/porn_gifs, /r/pornvids) along with myriad subreddits dedicated to specific sexual interests. However, /r/gonewild (GW) is by far the most popular NSFW subreddit. It features images of redditors in various stages of undress or engaging in sex acts. While GW's description ("Gonewild is a place for open-minded **Adult Redditors** to exchange their nude bodies for karma") suggests that any redditor might post, it is clear that the subreddit almost exclusively features images of women. GW's popularity on the site is noted and invoked regularly in interactions between (presumably) male and female redditors. When a photo featuring an attractive female holding a puppy, for example, reaches the top of /r/all, invariably multiple redditors will make a comment about looking to see whether the user posts to GoneWild or hoping that she will if she hasn't already (for example, DingoQueen, 2014). Recently, a woman posted images of her dad's work as a snowplow driver in Minnesota to /r/pics. The wildly popular post turned into a mini-AMA where redditors asked the OP about her dad's daily routine and discussed various aspects of driving in winter weather (pertnear, 2014). At one point, someone mentioned that the OP was 30, to which one of the replies was, "You're 30? /r/gonewild?" This behavior is so common that it spawned a bot, /u/GoneWildResearcher, that redditors can summon to automatically look for postings to various NSFW subreddits made by a particular user. In the thread introducing the bot, redditors discussed the possible ethical and privacy violations that the bot could pose (BearkeKhan, 2014). As one redditor noted sarcastically, "It was about time someone mechanized sexual harassment on reddit," to which the bot's author responded, "It's not meant to mechanize sexual assault and more meant to view it. Plus it could be used to stop idiots saying someone must have a gonewild post." While the author of the bot suggests that its intention is to highlight and perhaps discourage these kinds of GW questions (as bot queries would post to a subreddit set up for this purpose), this claim is dubious at best. If the reddit community were really concerned about this kind of harassment, reddit would create stricter moderation policies or actively enforce the ones it already has in place. Instead, the community seems passively to accept that women will often receive public comments and private messages about their appearance and sexual desirability/availability. And again, the request for GW images may not seem out of line for some "nice guy" redditors. After all, /r/gonewild is a space to which redditors can choose to submit their own pictures, thus reinforcing the idea that the women on the receiving end of these kinds of comments are active agents—even if the question itself seems merely another version of "Tits or GTFO."[7]

At the same time, a vocal segment of the reddit population suggests that GW contributors are "attention whores" and practice a fair amount of slut shaming regarding a woman's choice to post to the subreddit. Image macros appear in /r/AdviceAnimals with some frequency that compare the GW posters to crack addicts,[8] or hypocrites who are merely looking for attention from men they would not consider dating. For example, a Suddenly Clarity Clarence image macro reads, "Gonewild is a place where women beg for sexual attention from men they would never consider sleeping with" (uniquelyunqualified, 2014). The reactions from the community to these kinds of threads are mixed, with some recognizing that it constitutes slut shaming, others suggesting that the posters may have a sexual interest in exhibitionism, and still others arguing that GW posting stems from a deep-seated sense of insecurity and desire for attention. Or, as a commenter noted, "[They post] because they want to know men want to fuck them." The dual love-hate relationship with

GW is also a point of discussion, as one commenter noted, "I don't get reddit. Do they hate gonewild, or do they love it?" to which another replied, "They love it, but want people to feel worse about themselves than they do."

This kind of chatter suggests deep-seated insecurities that some redditors may feel about the opposite sex and themselves and a misogynistic view of women as objects existing solely for male discussion and pleasure. Furthermore, any privacy violations that occur as a result of GW posts being redistributed to other networks without the owner's permission are normalized, with the victim shouldering most of the blame as "she should have known better" (1to34, 2013). As one journalist notes, "… The pervasive idea is that as a female redditor you should only contribute to a conversation if you show your boobs, but once you do, you're a pathetic attention whore. The only appropriate place for women on Reddit is where they're naked and quiet, not trying to draw too much attention to themselves" (Dunn, 2013). And, as a recent GW poster who was called out for being an "attention whore" by another redditor noted, "Reactions like his are very frequent, and I believe it's due to resentment of the fact that women's bodies are desirable, and in being desirable, hold some inherent power. Men do lots of things to achieve penis-contact with women. So in his eyes, I am blatantly flaunting my 'power' and rising in status because of it" (AlleaGirl, 2014). Thus, a woman's agency, ownership of her own body, and image take a backseat to the pleasure she may provide for others on the site. It also suggests a double standard; a man can post to reddit, reap karma, and enjoy, but if a woman is posting to GW, she is doing so merely for the attention.

With the exception of GW, most NSFW subreddits feature material submitted by straight males for consumption by straight males. Most of this content appears to be created by professional porn actors or amateurs who willingly distribute it—whether or not they intended it to appear on reddit is another question. There are other subreddits, however, that feature content which may never have been meant to be made public or was only intended for a restricted audience or was obtained surreptitiously. The most famous example of this is the now-banned /r/creepshots, a subreddit that featured images of women in public taken without their knowledge. While Creepshots was shuttered by reddit administrators in 2012, content quickly moved to other subreddits such as /r/CandidFashionPolice (CFP), which still exists as of this writing (Alfonso, 2014). Despite CFP's tagline, "In this subreddit people post candid photos of women and then we judge their fashion choices similar to TLC's what not to wear and E!'s FashionPolice," and the ostensibly "fashion-oriented" titles that many of the posts feature, the subreddit is clearly carrying on the Creepshot tradition. The highest-rated posting (1402 points), for example, reads, "Girl, you need new sneakers," and features a picture of a woman from behind lying on the grass wearing short-shorts (JackRig95, 2013). Clearly, the photographer's intention was not to critique her poor shoe choice, as the sneakers are hardly visible in the shot, but to capture and share an image of a young woman without her knowledge for the sexual pleasure of others. The comments on the photo suggest this as well, with one redditor posting, "I have never seen shorts being sucked so far up someone's ass like this," and another responding, "It's a thing of beauty." The images on the subreddit are almost entirely from behind or from angles that would suggest the subject has no idea she is being photographed. Other subreddits are even more "subtle" regarding their intention. For example, both /r/girlsinyogapants and /r/YogaPants feature warnings suggesting that no creepshots/voyeuristic photos are allowed. And, yes, some of the photos seem to be self-shots (selfies), which were most likely posted to Facebook or Instagram. However, a larger percentage of them are clearly images taken from behind, as the "point" of /r/YogaPants and /r/girlsinyogapants is to admire the subject's backside. These subreddits trade on providing a sexual thrill for observers who know that the women pictured have no idea they are being photographed or that the images will end up being posted to and discussed on reddit. And, while some of them have been banned, new ones continually appear. Reddit administrators rarely step in to police content or adjudicate disputes, instead leaving it to the community and subreddit moderators. This tacit acceptance means spaces like /r/creepshots will continue to populate the fringes of the reddit community.

/r/MensRights, /r/TheRedPill, and /r/seduction

Anti-feminist communities have a strong presence on reddit. /r/MensRights (MR) is a medium-sized community (about 89,000 subscribers as of this writing) that acts as a catchall subreddit for anti-feminist men's activists.[9] The subreddit's FAQ suggests that it serves as a space for "… honest discourse in regards to male issues—including but not limited to custody, alimony, reproductive health and rights, and education. MR is a subreddit consisting of both men and women who believe that there is serious discrimination against men inherent in western societies" ("faq—MensRights," 2014). Postings in the subreddit range from discussions of false rape claims, the fact that men are the targets of violence more often than women are, how child support agreements are unfair to fathers, and how men are subject to unrealistic body standards. As a feminist, trawling through MR is difficult, not so much because I disagree with all of the evidence they provide or the opinions they espouse but because the conversation frequently devolves into blaming feminism for the problems that they face. For example, many Men's Rights Activists (MRAs) view feminism as creating an environment in which an overwhelming number of men are falsely accused of rape.[10] Instead of understanding how they themselves are negatively impacted by a complex matrix of gender, class, and racial politics, they tend to place blame at the feet of women and, particularly, feminists (for more on the vagaries of the men's rights movement, see Kimmel, 2013). Thus, even discussions of important issues (such as the reality of female-on-male domestic violence and sexual assault) are subjected to dripping sarcasm with statements such as, "Didn't you know? Only men are capable of violence. Wymyn would never do such a thing" frequently appearing (Down-n-Dirty, 2013). Not surprisingly, Southern Poverty Law Center included the subreddit in its 2012 *Intelligence Report* of misogynistic spaces online (Southern Poverty Law Center, 2012).

/r/TheRedPill (TRP) takes its name from a scene in *The Matrix*, in which Morpheus offers protagonist Neo the option to take a red pill to see "the real world." Unlike MR, this subreddit channels resentment against women into a sexual strategy by suggesting that they secretly wish to be dominated by men and that feminism is hurting this "natural order." Drawing on a terrible application of evolutionary psychology, TRP suggests that men are lacking masculine role models and that women have been fooled into thinking that what they'd like is equality—thus, TRPers suggest that men master "game" rather than blaming feminism for their troubles (pk_atheist, 2012). In this case, "game" is a set of rules for interacting with women gleaned from a collection of pickup artists that *Rolling Stone* reporter Neil Strauss (2005) profiled.[11] /r/seduction focuses specifically on picking up women for sexual relationships, whereas TRP suggests the "game" should be applied to all areas of life—including long-term relationships, work, and so forth. Both subreddits suggest that women secretly hope for domineering men, even if they say otherwise. It is hard not to see how this might work its way into an ethically questionable area regarding consent.

Why have these communities coalesced on the reddit platform? I suspect several reasons. First are the demographics of reddit and its tendency to reinforce ideals of geek masculinity. Second is reddit's administrators' extreme unwillingness to intervene when it comes to content moderation, allowing communities like this to flourish. Given that even extremely disturbing subreddits such as /r/niggers (which was finally banned in 2013 after much drama) or /r/PicsOfDeadKids (which is still public and active) have been present on the site for some time, the idea that reddit admins would take issue with "merely" misogynistic subreddits such as MR and TRP is unlikely. Third is the reality that reddit merely reflects the internet (and culture) as a whole. These are real impulses in real people that are merely made more public in online spaces—and it reflects a culture of white male privilege that remains relatively unquestioned. Sociologist Michael Kimmel (2013) refers to the anger of the men's rights movement as part of a larger culture of "aggrieved entitlement" that dominates the thinking of a certain segment of American white males. His observations seem apt—the MR and TRP communities project their frustration about their circumstances on to the backs of feminism and feminists.

As with the GoneWild subreddit discussed above, MR and TRP tend to "leak out" into more mainstream areas of reddit, meaning that commentary on other subreddits may reflect the perspectives advocated within these niche communities. This is also the case with other unsavory subreddits, such as /r/niggers. As *The Atlantic*'s Bridget Todd argues, "While the subreddit's postings were unquestionably racist and offensive, what was really disturbing about r/niggers was the way the group's commentary and subscribers seeped into the broader Reddit community at large. It became a launching pad for excursions into the rest of Reddit. This particular dark corner of the web was never merely content to stay in its corner; its members ventured out" (Todd, 2013). The misogynistic views of TRP and MR do not simply stay put in those subreddits; they become part of the larger reddit culture—informing the ways in which women are discussed and treated on the rest of the site. For example, a recent posting to the /r/videos default subreddit of an anti-feminist speaker being "shut down" at a Canadian university led to an intense discussion of feminism, most of which characterized feminists as radical, man-hating individuals interested in promoting female superiority (hitbart000, 2014). It also becomes clear that these views are promoted when looking at something as relatively banal as image macros and memes.

Scumbag Stacys, Good Girl Ginas, Overly Attached Girlfriends, Redditor Wives, and College Liberals

A number of image macros perpetuate a discourse that reinforces Nussbaum's (2010) idea of a "denial of autonomy" when it comes to how women are objectified online. While these memes may be distributed across other networks, they are popular features of reddit's AdviceAnimals subreddit. Each presents a stereotypical view of women that almost always is told from the perspective of a man who is often a romantic partner. Many feature images that are remixed versions of stock photography images or are images "borrowed" from social networking or other public web sites.

"Scumbag Stacy" features a young white woman in a messy bedroom (presumably a dorm room), provocatively clothed in a tank top, underwear, and baseball cap turned to the side, and who is showing a belly button piercing. Her facial expression is something between a sneer and a smirk and her face is covered in heavy makeup. In contrast to the male version of this image macro, "Scumbag Steve," Stacy is clearly supposed to be viewed as a sexual object (Steve is fully clothed, and cropped from the chest up, whereas Stacy is shown in a medium shot with a bare midriff). And, indeed, the real "Scumbag Stacy" (Amber Stratton) was the winner of College Humor's 2007 Hottest Girls on Campus contest (Triple Zed, 2013). Stacy macros typically feature a short story of deplorable behavior indicating her apparent selfishness, narcissism, unwarranted resentment, cluelessness, or adulterous behavior. Often this behavior is displayed in the context of a romantic relationship she has/had with the person submitting the image macro. For example, the first line of a popular Stacy meme reads, "Breaks up with you and tells you not to talk to her," with the second line reading, "'You were supposed to fight for me!'" (project8an, 2013). Another example reads, "Goes to CSA [Child Support Agency] because she wants more money," followed by, "Mad at me because they determined I was paying too much" (lukeis2cool, 2013). Still another reads, "Gets mad because you hooked up with a girl a year ago before you even met," finishing with, "Texts ex on a daily basis and sees nothing wrong with it" (LavaLampJuice, 2013). Responses to Stacy tend to side with the poster, suggesting that the woman in question is, indeed, a "psycho bitch" or that her actions are those of all women, which is probably the fault of images they see in romantic comedies, or that the poster should "Lawyer up, delete Facebook, and hit the gym" (a phrase so often repeated in threads about relationships that it has been remixed in multiple nonsensical permutations—for example, "Hit the lawyer, delete the gym, and Facebook up"). While "Scumbag Steve" macros offer equally odious examples of men acting selfishly, the general tenor of those conversations focuses on how that particular person is a jerk rather than making a categorical assessment of men in general. Not surprisingly, Amber Stratton has filed cease-and-desist orders against sites such as *Know*

Your Meme that continue to use her image without her permission; however, as of this writing, her image still appears regularly on /r/AdviceAnimals and /r/all (Brad, 2012b).

"Good Girl Gina" (GGG) is the polar opposite of Stacy. Whereas Stacy is imagined as an out-of-control harpy whose histrionic antics are unreasonable and completely unprovoked, Gina is compliant, sexually available, and considerate to a fault. In contrast to Stacy's appearance, Gina features a close-up shot of a young white woman wearing natural-looking makeup, smiling sweetly with her hand on her slightly tilted chin. Gina is the female counterpart of the "Good Guy Greg," an image macro used to report considerate things that a "good guy" might do for someone else. Greg is usually doing things like cleaning up after himself at parties or seeding/sharing torrents or paying back more money than he had borrowed (aux, 2012). Gina, however, offers to make dinner, is a willing sexual partner, approves of her mate's lifestyle (gaming, drug use, etc.), and doesn't "friendzone" men. A redditor in 2013 offered a comprehensive analysis of Gina memes, finding that they frequently relied on objectification, her sexual availability, and her distinctiveness from other women. LaTeX_fetish summarized thusly:

> If someone wants to be a Good Girl, then reddit already has it figured out. A Good Girl is an object to be lusted after. A Good Girl makes sure you're sexually satisfied, either by her or someone else. A Good Girl defies stereotypes, unless they play into your desires, like when she cooks for you. A Good Girl plays your video games and watches your movies, and she'll bring you food and drinks and drugs, but a Good Girl won't talk about any of those things, because she is a Good Girl. And a Good Girl keeps quiet and doesn't rock the boat. (LaTeX_fetish, 2012)

Unlike Stacy, who is far too sexually available, Gina is sexually adventurous only when it comes to the person with whom she's in a relationship. Gina is not a tease—she will have sex with you when requested and happily watch as you play video games. This sort of virgin/whore stereotype suggests that at least a portion of the reddit community subscribes to a vision of womanhood that denies any real autonomy for the person in question.[12]

The "Overly Attached Girlfriend" (OAG) image macro has a different genesis from that of either Stacy or Gina. It is inspired by a YouTube video in which a young woman (Laina Morris) sings a parody version of Justin Bieber's "Boyfriend," replacing the lyrics with slightly off-key, clingy, and stalker-ish ones ("If I was your girlfriend/I'd never let you leave/without a small recording device/taped under your sleeve," Laina, 2012). A still of the young white woman staring wide-eyed into the camera at the beginning of this video serves as the template for this macro (Brad, 2013e). The woman appears to be in a college dorm room, with a cork board filled with pictures of herself and her friends in the background, dressed simply in a blue T-shirt wearing makeup that looks relatively natural but appears somewhat off kilter (orange lips and darkened eyes). In the video, her facial expression moves from creepily adoring to deadpan, with her wide eyes remaining one consistent feature. OAG features stories of romance gone wrong—projecting the girlfriend as a clingy mess who is far more invested in the relationship than the macro's creator is. For example, one OAG was posted under the title, "I told my girlfriend that I didn't want to have sex because I was too tired. So right before we got it on she said this," with the image macro reading, first, "If you fall asleep" then, "You will wake up a father" (Bobby_Ooo, 2012). Or in a post titled, "This just came out of a coworkers [sic] mouth," the OAG reads, "I'd show you a picture of him. But I don't have any of him awake" (Ireallylikepbr, 2014). Another, titled "My girlfriend pulled this on me today," featured OAG with the lines, "Sends you her Amazon wish list" and then, "All engagement rings" (joserskid, 2014). OAG macros are intended to be humorous, but the humor is based on exaggerations of stereotypical images of women as needy and unreasonable and the likelihood of stalking/being suspicious of current or former romantic partners. This appears despite evidence that while both men and woman in the US are equally likely to experience harassment, women are more likely to be the victims of stalking behavior (Baum, Catalano, Rand, & Rose, 2009). Of note is how Laina Morris has embraced the Overly Attached Girlfriend image. She conducted an AMA for the AdviceAnimals subreddit (LainaOAG, 2012), later appearing in a Samsung commercial and acting as a red-carpet host for the 2012 American Music Awards (Marin, 2012). There is also a version of OAG that suggests men can also be clingy, "Overly Attached Boyfriend," but it has never experienced much popularity. Responding to one version of the boyfriend image macro, a redditor wrote, "This isn't going to work, just like it didn't work the last few times. It's just like don glover once said. You never hear women complaining about crazy boyfriends, because if you have a crazy boyfriend you're going to die" (supdunez, 2013).[13]

"Redditor's Wife" (RW) (also known as "Internet Husband") is a modification of a stock photography image featuring a man staring into a computer with his (what appears to be) long-suffering wife or girlfriend slightly out of focus looking down in the background (VanManner, 2013). This image macro plays on the idea of redditors coded as male (discussed in Chapter 3), and it usually references a current meme that is popular on the site that would make little sense to those outside of the community. Presumably the first line represents something the redditor's wife says, and the second line represents something that the husband does or says in response. Often the memes suggest that the RW is sexually frustrated. For example, one RW image macro referenced a Confession Bear macro that confessed to using the euphemism "take a nap" for masturbation. Another poster responded with an RW that read, "I asked him if he wanted to have sex tonight. / He said 'no thanks I took a nap earlier'" (jrivers242, 2014). Other RW macros reference particular events that occurred on the site. After Bill Gates participated in an AMA and received Reddit Gold from many different community members, the RW macro "redditor's wife fooled three days before valentine [sic] day" was posted. It read, "I asked him why I saw a charge for gold on my credit card. He said 'It's for Bill Gates'" (Mcodray, 2013). Still others reference common, in-group phrases regularly seen on the site, such as a particularly grotesque one that reads, "I said I had a miscarriage / 'OP didn't deliver'" (MuscleT, 2012). This RW struck a nerve with several posters, who suggested that the entire premise was not humorous, having themselves experienced the pain of miscarrying. The image of the "Redditor's Wife" reinscribes the problematic and false myth of there being "no girls on the internet"[14] and perpetuates the unfortunate idea that reddit is somehow not for women.

The College Liberal (CL) image macro directly critiques progressive ideals, particularly in regard to feminism. The macro features an image of a white woman (or possibly a man—but most often the poster uses CL to comment on feminist issues) with dreadlocks wearing a knitted cap, orange hoodie, and glasses with a self-satisfied look on her face. According to *Know Your Meme*, the image was likely taken in 2005 but did not reach popularity on reddit as an image macro until late 2011 (Brad, 2013a). A prototypical example of the CL image macro might read, "Argues how gender roles don't exist / 'What? I'm not asking him out [.] That's his job'" (AquilaGlobumAncora, 2013). Another example, titled "I don't hate female bosses, just my female boss," reads, "Claims sexist hiring and promotions hurt business / before announcing that she will only be promoting women to C-level positions" (readeranon, 2013). This led to a number of commenters suggesting that the poster should file suit against his boss and that she was probably using the "feminist" definition of sexist. Others suggested that the meme should be renamed to "straw feminist," as they knew no one who would actually suggest these kinds of things, with others suggesting that as feminists, they were ashamed by this woman's actions.

All of these image macros work to support an image of women as sexually desirable, but only when they are compliant and subsuming their own needs to those of their partners. They also suggest that women are not to be trusted, as they are likely to cheat or otherwise take advantage of their "nice guy" partners. And they imply that feminists are likely to be humorless, shrill harpies who randomly misapply these ideals to situations in an effort to shame or insult the men in their lives. Women who deviate from the "Good Girl Gina" image must be shamed because they have broken unstated rules around what constitutes appropriate femininity. Importantly, all of these image macros present women from the perspective of men—although the gender of the creators of these remains unknown, in many cases the stories that the OP shares suggest they are heterosexual men discussing their wives, girlfriends, ex-lovers, or female coworkers. And often the goal of these image macros is to humor others on the site—to engage in the same kind of play discussed in the last chapter. But this play takes on a more pointed and passive-aggressive form, refashioning the whimsy of the image macro into a thinly disguised version of misogyny and resentment. Thus, the "carnival" here takes a dark turn beyond merely reveling in the grotesque to challenge dominant power structures to glorifying these power structures themselves.

/r/ShitRedditSays (SRS) and its contrarians

Along with the problematic gender politics mentioned above, much of the discourse on reddit trades in casual racism, ableism, transphobia, and homophobia. While this is not exclusive to reddit, as Lisa Nakamura's (Nakamura, 2002, 2007; Nakamura & Chow-White, 2012) extensive work on race and identity in online culture demonstrates, the combination of the community's valorization of geek masculinity and an "anything goes" policy when it comes to content moderation tends to create a space in which this kind of discourse remains mostly unchallenged. One subreddit, ShitRedditSays (SRS), attempts to expose this reality by linking to and discussing comments and posts deemed offensive. SRS's description reads, "Have you recently read an upvoted Reddit comment that was bigoted, creepy, misogynistic, transphobic, racist, homophobic, or just reeking of unexamined, toxic privilege? Of course you have! Post it here." SRS (/r/ShitRedditSays), also known as SRS "prime," functions as a "circlejerk" (much like the popular /r/circlejerk, which pokes fun at the stereotypical interactions you might see on the default subreddits discussed in

Chapter 3), meaning that it is a safe space for the SRS community to play and generally decompress from the toxic environment that characterizes some areas of reddit. The subreddit's FAQ notes, "It's our space and we don't have to make room for people who don't 'get it'. More to the point, SRS is a place for those who already know why something might be considered offensive; not for those who wish to find out why" (ArchangelleSamaelle, 2013). SRS members take issue with the idea that they should somehow make their message more palatable by changing the tone of their message, often referred to as a "tone argument" ("Tone Argument," 2014). Other areas of SRS (called the FemPire) offer a mirror of well-trafficked areas of reddit with an SRS bent. This includes SRS versions of popular subreddits such as /r/SRSFunny, /r/SRSGaming, and /r/SRSTechnology. All parts of the SRS FemPire are heavily moderated, with users being banned for "posts that are bigoted, creepy, misogynistic, transphobic, unsettling, racist, homophobic, or just reeking of unexamined, toxic privilege" ("SRSTechnology," 2014). It is important to note that given the discussion of reddit's culture of play and humor (discussed in the last chapter), using slurs "ironically" or "satirically" is also against SRS rules ("ShitRedditSays," 2014). This choice stands in stark contrast to a kind of "hipster racism" and sexism that pervade the discourse on many other subreddits.

At the very beginning of this project, I stumbled across references to SRS and decided that I wanted to include their voices in the discussion of the more problematic aspects of reddit culture. However, I knew that the SRS community was often under attack from various groups, both inside and outside reddit, and I was not sure how willing they would be to talk with me. After I posted a question to the /r/circlebroke subreddit regarding SRS's origins, one of the SRS moderators responded. After I offered proof of my identity through my Twitter account, a number of the moderators graciously agreed to discuss SRS and its tense relationship with the larger reddit community.

SRS is a polarizing entity. Entire subreddits are dedicated to deconstructing SRS goings-on, with endless discussions of its (supposed) role as a "downvote brigade." SRS members are accused of logging in with secondary accounts just to downvote material they find offensive or downvoting entire submission histories of certain individuals. SRS moderators deny this, suggesting in fact that part of their goal is to make sure problematic content across reddit is highlighted (thus their choice to highlight and discuss that content on SRS prime) rather than being hidden by the site's sorting algorithm.

Site admins have also denied that SRS breaks rules at any higher rate than do other subreddits' members, and that if anyone does doxx or otherwise harass other redditors, those accounts are summarily banned (alienth, 2013). But the aura surrounding SRS is rather astonishing. In almost any thread discussing gender or racial politics, at some point someone will mention SRS and this will spawn a long diatribe about the nature of SRS, how unwilling its members are to engage in "rational" discourse with the rest of reddit, and why they remain on the site at all. For example, a recent /r/AskReddit thread asked for others to share interesting, secret things about the site. One of the most upvoted and gilded comments (930 points as of this writing) provocatively mentioned that "a certain subreddit" was not shut down despite breaking the rules of reddit:

> That a Certain subreddit receives diplomatic immunity from Reddit's mods despite repeatedly breaking Reddit's code of conduct, Witch hunting, Doxxing and Brigading other members on a regular basis.
>
> As to why this is allowed is a complete mystery!
>
> Anyone care to guess which sub I am referring to? (Sirinon, 2014)

Someone responded by asking whether the commenter meant SRS, which spawned a long series of discussions about SRS and its supposed cozy relationship with the site administrators. Others continued to speculate, with one redditor writing, "It's a known fact that 1/3 of reddit admins also moonlight as SRS archangelles (mods)," with still other redditors chiming in with claims that SRS doxxed Violentacrez (which is not true—Violentacrez self-identified at several reddit Meetups; see Chen, 2012) and encouraged someone else to commit suicide (also not true—see ArchangelleDworkin, 2012; Coscarelli, 2012). Other threads, many of which are archived in SRS's own repository for these stories, /r/SRSMythos, suggest that one of the SRS moderators is dating one of the reddit administrators. Still other threads suggest that SRS is a bigoted, hateful group of shrill "feminazis" and "white knights"[15] that resort to cynicism and sarcasm instead of engaging in "rational" debate around gender and racial politics.

Redditors' perspective on SRS's origins are equally conspiratorial and bizarre. The most common story was echoed recently in a thread on /r/FeMRAdebates, a small subreddit dedicated to discussions around feminism and men's rights, by one redditor:

> I know the basic history of SRS so take it for what you will:
>
> Users from "The Something Awful Forums" went into Reddit and formulated a thread in an effort to mock and satirize the most outrageous things said in the name of Social Justice on Reddit itself. That was what the true meaning of SRS was about.

> Then, extreme Social Justice Warriors mentioned glimpsed the posts and comments then began taking it seriously. Next thing you know, they flocked en masse to the thread and ensnared everything. Now, it's all about "This is my idea of social justice. Everyone else can fuck off!".
>
> In summary: What was meant as a joke thread intending to mock extreme things said in the name of justice on Reddit turned into the very thing it mocked. (Samwalter, 2014)

In other words, some view SRS as a place that was intended to parody radical feminism, which later was "taken over" by the current crop of SRSers. According to SRS's own history, some individuals from the Something Awful forums[16] did end up becoming moderators, but the beginnings of SRS can be traced to a posting in the now-defunct /r/LadyBashing. Someone suggested that what was needed was a very strong "shoot first, ask questions later" moderation policy, because by this point /r/LadyBashing had been overrun by Men's Rights Activists (ArchangelleStrudelle, 2012). SRS was the result. The subscriber base grew significantly after Skepchick linked to the subreddit in a 2011 blog posting damning the /r/atheism community on reddit for being sexist but ending with the postscript "… r/shitredditsays makes Reddit worthwhile" (Watson, 2011). Since then, SRS has been involved in several attempts to rid reddit of child pornography and creepshots (plasmatron7, 2011).

Part of the intensity of feelings that some redditors have toward SRS is based in its involvement in getting a number of questionable subreddits and accounts banned for sharing illegal, unethical, and otherwise objectionable material. SRS was instrumental in Pedogeddon, a campaign to, according to one journalist, "paint Reddit … as a den of child pornography—and free-speech-loving redditors as complicit pawns in its spread" (Morris, 2012a), which eventually led to the shutdown of /r/jailbait and the doxxing of infamous moderator Violentacrez (Chen, 2012). As a result, SRS has earned the ire of some other redditors and led to the spawning of a number of anti-SRS subreddits (for example, /r/antisrs, /r/AngryBrds, and /r/SRSSucks). Each of these subreddits has a slightly different tone, but all are critical of SRS's approach to social justice issues or the entire idea of feminism or anti-racism or both. The FAQ on /r/antisrs suggests that it is a space for those individuals interested in social justice issues but who are not happy with the tactics that SRS uses ("/r/antiSRS FAQ," 2014). However, after looking at a number of posts in antisrs, it is hard to see much representation of individuals who are actually socially justice minded. /r/AngryBrds, a play on both the popular mobile game *Angry Birds* and the SRS mascot of a bird (called BRD), features a header of a number of outspoken individuals who have drawn particular ire from the men's rights activists and some within the geek community, including Anita Sarkeesian, Adria Richards, Rebecca Watson, and a Canadian feminist who tangled with MRA activists after a talk sponsored by Equality Canada (a men's rights group) at the University of Toronto (Romano, 2013).[17] SRSSucks's purpose is, as their sidebar notes, "to make fun of what the users of SRS believe about reddit." The subreddit's FAQ suggests they take issue with SRS for a number of reasons, including their tendency to use sarcasm and mockery, and their tendency to "… frequently take their opponents out of context, so you have a sense that you need to rhetorically posture yourself in order to avoid being distorted" (MittRomneysCampaign, 2013). Like other anti-SRS subreddits, site moderators have altered the default CSS file for their community and swapped out the default upvote/downvote arrows with two icons—a silhouette of a man (hover text: "Manly men."), which represents an upvote, and a silhouette of the SRS bird mascot (hover text: "BRDS"), which acts as a downvote.

SRS as counterperformance and transgressive play

We could simply view misogyny and racism on reddit as just how things are on the net and leave it at that. But as Terri Senft argues, "One way to reassert the political importance of the personal is to move away from a preoccupation with *why* something is so, toward a more rigorous analysis of *how* certain behaviors network into a naturalized version of 'the way it is'" (Senft, 2008, p. 118). Part of what is threatening about SRS's actions is that they challenge the unstated assumptions that underlie much of reddit's discourse. Members question the *doxa* of the reddit community, or "the fundamental assumptions and categories that shape intellectual thought in a particular time and place which are generally not available to conscious awareness of the participants" (Swartz, 1997, p. 232). In Pierre Bourdieu's formulation, *doxa* refers to the ways in which fields are legitimated and how knowledge within intellectual traditions becomes naturalized, or more broadly, "the natural and social world appears as self-evident" (Bourdieu, 1977, p. 164). For Bourdieu, conflict between the established orthodoxy and the upstart heterodoxy is paradigmatic of all cultural, economic, and intellectual fields. Those who challenge the orthodoxy may do so through a strategy of *succession*, by trying to access positions of power, or by a strategy of *subversion*, which is "a more or less radical rupture with the dominant group by challenging its legitimacy to define the standards of the field" (Swartz, 1997, p. 125). It appears that the SRS community is less interested in accessing a position of power within the larger reddit community than with calling into question the norms that govern many interactions on the site. By subverting the strategies of play that those within the dominant positions of reddit employ, SRS directly challenges ways of knowing and interacting that rely on misogynistic, racist, and homophobic discourse.

I mentioned the performative nature of reddit, and specifically how the community's use of play reinforces community connections and creates a shared sense of history. As Richard Schechner notes, "The force of the performance is in the very specific relationship between performers and those-for-whom-the-performance-exists" (Schechner, 1985, p. 6). The reddit community likely resists the idea that they are the "audience" for whom SRS is performing, instead viewing themselves as the target of the performance. SRS is, in fact, engaging in a kind of counterperformance (Alexander, 2004), reinterpreting reddit's forms of "play" to offer a broader political critique of the site.

This is done in a few different ways. Most noticeably is the inversion/subversion of the rules of play around karma. Rather than employing the typical upvoting and downvoting mechanisms, SRS moderators have altered the subreddit's CSS file to allow only downvotes on the SRS prime home page. In this case, the downvote arrow actually acts as an upvote for the purposes of karma. The most recent posts receive the largest number of "downvotes" then at the top of the subreddit's home page. The actual comment threads within SRS feature both downvoting and upvoting arrows, but again, their functions are reversed. By this clever inversion of reddit's voting system (making upvotes downvotes and vice versa), SRS moderators say that they are interested in *highlighting* the "shit reddit says," thus ensuring that this stuff still remains *visible* instead of hiding it underneath a sea of downvotes. Their point is that the toxic stuff some redditors are spewing deserves to be seen by the community as a whole. It also acts as a buffer for those who are unfamiliar with the subreddit's voting mechanism and may visit from the default or anti-SRS subreddits to brigade or blindly downvote material they disagree with. In this case, downvotes would be recorded as upvotes for the SRS contributor. In an interview with /r/subredditoftheday (a subreddit dedicated to highlighting new and interesting subreddits of which redditors may not be aware), one SRS moderator put it this way:

> First off, if SRS consistently downvoted the posts it links to, it would sweep the bigotry and hate that a lot of people have to actually deal with under the rug, which is totally counterproductive to the general groove of SRS. If everything we link to gets −50 votes or whatever, it makes the fact that Reddit is pretty overtly hostile to anyone who is not a white straight male 1%er seem like less of a problem than it actually is.
>
> More importantly, I really am not comfortable declaring Reddit to be the locus of social justice activism and progressive liberalism that a lot of our critics seem to think it is. If we really want to tell ourselves that mashing the down arrow on posts that openly declare their bigotry is going to Solve The Problems then honestly, we're deluding ourselves ...
>
> Quite simply, the reason why we do not downvote brigade is because we seek to preserve the shit that has been said. Immortalize it, so to speak. To hold people accountable for the shit they've said ... Another reason why we don't downvote brigade is because that de-legitimizes us in the minds of other redditors and reinforces a mob mentality. (jmk4422, 2011)

In addition, SRS's actions also implicitly critique the notion of karma as a representation of "value" and visibility within the reddit community. If an offensive comment is highly upvoted, it will also increase in visibility (as most people are likely to browse reddit without changing the sorting order when viewing discussion threads), which implies a kind of validation of the said comment or perspective to other redditors. If SRS subscribers then enter into the thread and question the value of the perspective expressed by the OP, they tend to be downvoted, thus burying their critique and making it less visible to later visitors to the page, as postings with negative karma below a certain threshold are both grayed out on the page's loading and must be "expanded" by clicking on the plus sign next to the comment. The reversal of the downvotes and upvotes in SRS prime therefore serves to demonstrate SRS's perspective of the meaninglessness of karma, to prepare SRS subscribers for the inevitable downvotes they will receive outside the FemPire, and to challenge the notion that the reddit community is actually progressive or interested in creating an egalitarian, deliberative space.

SRS also subverts and plays with the dominant discourse of reddit. Often threads within SRS prime will perform a kind of commutation test (Barthes, 1977) and remix actual sexist or racist opinions expressed on the rest of reddit, replacing the target of the racist or sexist remark with the default, stereotypically white male redditor. While this kind of play is accepted, or at least tolerated, in /r/circlejerk, most likely because all aspects of reddit culture are parodied, SRS's approach is derided at best as nonsensical and at worst as bigoted. SRS members are considered by some on reddit as a kind of "thought police" that is trying to impinge on the rest of the community's free speech (or, as it is jokingly referred to in some of the meta-subreddits, "freeze peach"). To further ensure that the SRS prime remains a circlejerk and the rest of the FemPire remains a "safe space," moderators are quick to ban individuals who are there to troll or critique SRS's approach.

Insider vs. outsider critiques

SRS is threatening to at least a substantial portion of the reddit community for several reasons. Returning to the idea of geek masculinity, much of the implicit critique of SRS has something to do with the fact that many within geek culture view themselves as subject to hegemonic standards of masculinity with which they do not comply. Therefore, they may, as Zeynep Tufekci (2014) astutely argues, see themselves as quintessential outsiders—not recognizing the vast amount of social and cultural capital to which they actually have access. Their privilege remains relatively unrecognized and unstated, especially in the echo chamber of reddit, where many individuals with a similar background (male, white, college educated, STEM oriented, etc.) coalesce.

Although the SRS's FAQ compares the subreddit to /r/circlejerk, their approach to subverting the dominant culture of reddit is different. As discussed in Chapter 4, /r/circlejerk creates a Dadaist space of play where every aspect of reddit is fair game for mockery, from the most banal to the most serious. SRS's critique is far more targeted, highlighting the unspoken assumptions that redditors make about issues of class, race, gender, and sexuality. When I asked SRS moderators about why /r/circlejerk seems tolerated (as evidenced by its regular appearance on /r/all) and SRS is reviled, one suggested that /r/circlejerk's parodic nature is viewed as an extension of other forms of play on reddit, whereas SRS's critique is much more pointed and direct: "Circlejerk's critique of the hivemind is pretty playful and non-threatening whereas ours is … not. From the beginning we've made no effort to avoid antagonizing people who aren't receptive to what we have to say, in fact we've welcomed it. We've publicly shit all over reddit's brand in the press on multiple occasions. We took away their jailbait and their creepshots. We got their favorite Starcraft players fired or suspended." Thus, /r/circlejerk's form of play somehow still falls within the carnivalesque atmosphere of reddit (perhaps annoying but ultimately harmless), whereas SRS's play questions the fundamental nature of the carnivalesque itself—and suggests that it is implicitly reinforcing dominant power structures.

Perhaps unsurprisingly, many redditors outside SRS seem to take umbrage at their direct and unrelenting approach to critiquing the community, suggesting instead that they would rather have a rational dialogue about these issues. As one redditor noted in the same AskReddit thread mentioned above,

> Circlejerking about racism, sexism, and whatever else doesn't solve anything. It just spreads [more] hatred than initially there, and they come off like assholes. You don't solve a problem by causing another problem of the same kind. It would be much more productive to actually have a civil conversation about why this or that is wrong. (Sirinon, 2014)

While this redditor would, perhaps, be open to an honest dialogue about issues of class, gender, race, and sexuality, not all are. And after having spent many, many hours on reddit, reading threads, writing field notes, and interviewing users, I can honestly say that these kinds of conversations *cannot* happen in the most popular subreddits—at least not as the community currently exists. This is because the large number of individuals involved makes separating out the strands of conversations happening all at once almost impossible and moderation extremely difficult. Both elements are imperative if these dialogues are to lead to thoughtful deliberation about these issues. More important, this kind of tone policing is a common tactic for those who do not really want to engage in these kinds of difficult dialogues to begin with, which, I am afraid, characterizes much of the reddit community. As one SRS moderator told me, dialogue with others on reddit was not productive, and, unfortunately, was emotionally damaging:

> Back when I first joined reddit I thought that discourse was the best way to convince people that their ideas were wrong. After all, I had come to my feminist conclusions through reading, discussion and general introspection. Surely everyone was the same way I was. Then I was told by several redditors on mensrights that I deserved it when I was raped. That was my lightbulb moment. Talking with these fuckers doesn't do shit.

Another moderator noted a similar problem when trying to engage in honest dialogue with many others outside SRS, writing,

> I used to believe that if people sat down and talked things out they'd "naturally" come to the conclusion to be nice and respectful to everyone, and learn to understand each other and everything would be hunky dory. I was learning that cognitive dissonance basically destroys this thanks to experiences I had leaving a religious cult and having

difficulty even figuring out where to start in trying to explain to my family and coming to the conclusion that *I shouldn't and it'd be harmful to even try*, but Reddit taught me just how bad the problem is, and that it goes far beyond religious cult members and permeates everything. (emphasis in original)

Both of these individuals felt that the emotional cost of engaging with reddit became too great a price to pay—especially given that many individuals were interested only in promoting their own "knowledge" of others' life experiences rather than actually listening or somehow having an honest, real dialogue.

And it does not help that SRS is viewed as being somehow "outside" the reddit community, despite their very real presence on reddit. An image macro that made the rounds on reddit suggested SRS comes to the site "just to bitch about it"—as if these individuals aren't "legitimate" members of the community ("Comes to Reddit Only to Bitch about It," n.d.). SRS becomes a boogeyman trotted out in any thread where people challenge the often oppressive, Othering views of gender, race, sexuality, and disability status that many redditors express. This happens despite attempts by at least some of the SRS moderators to suggest that they were simply fed up with getting downvoted when trying to actually engage with redditors on these tricky issues. And so SRS acts as a kind of "safe(r) space" where individuals can socialize and decompress without needing to confront the toxic culture that permeates many other subreddits. As one moderator said, "I had been heavily involved in the gaming community, and was tired of the constant onslaught of racist, sexist, and homophobic jokes/comments. SRS let me know that I wasn't alone. It was a space where I finally had people who thought that shit wasn't funny—and stuff they found funny, so did I … SRS let me know that my voice was allowed, and that I wasn't 'crazy.'" For most of the moderators I talked to, SRS served as an important outlet for expressing their very real frustrations with many aspects of reddit, and, by extension, geek culture more broadly.

However much parody, self-deprecation, and meta-talk characterize reddit's discourse (discussed in detail in Chapter 4), the community also has a serious problem with critiques that are perceived to come from the "outside." Compare, for example, the reception that actor William Shatner, famed for his geek-friendly role as Captain Kirk on *Star Trek*, received when he noted the toxic environment on reddit with the response that SRS typically receives. After he asked about why accounts that expressed racist, homophobic, and sexist opinions were not banned by moderators, he was not only upvoted but met with mostly intelligent comments about reddit's culture (williamshatner, 2013). A thoughtful conversation about the nature of reddit broke out on the posting, with a number of different redditors chiming in to express their opinions. Some suggested that part of the site's charm was its mixture of the profound and the profane—the latter underscoring and making the former more notable. Slight admonishments from reddit administrators to "remember the human" behind postings are also met with agreement and thoughtful commentary, with, of course, a few "fuck yous" thrown in that linger near the bottom of the thread because of downvotes (cupcake1713, 2014b). SRS, conversely, is met with outright hostility. So, too, are journalists such as Adrian Chen or critics such as Rebecca Watson who publically take issue with some elements of the site's culture. This kind of insider/outsider tension mirrors the realities of discourse in many other geek-friendly communities—for example, the charge that Anita Sarkeesian's "Tropes vs. Women" videos are irrelevant critiques because she is not a "gamer" or somehow does not belong to the "gaming community." Of course, tensions such as these, of who belongs and who does not, of authenticity, and of who is allowed to speak are common in all kinds of communities. But I think that the technological logics of reddit and the internet more broadly make these kinds of moments both more dramatic and more public.

What I find interesting about SRS and the FemPire more broadly is that the community purposefully exists within the space of reddit, hoping to annoy, to frustrate, and perhaps to serve as a reminder that many people are *not okay* with the kind of discourse that sometimes dominates reddit's default subreddits. But they do not move elsewhere—and do not create their own version of reddit to engage in this critique. Thus the carving out of a space where they can create their own forms of play—where the carnival nature of reddit is further inverted to critique the dominant discourse on which much of the larger community relies. SRS's moderators have expressed that there are other aspects of reddit that they find important. They are part of the community. And, as they said, even if they were to "move off" reddit, it's not as though this kind of speech isn't everywhere on the internet more broadly.

This may also be the reason SRS draws the ire of a portion of the reddit community in ways that subreddits trading in racist, sexist, and homophobic content do not. When I try puzzling out why SRS remains the frequent target of public speculation regarding its presence on the web site, whereas subreddits such as /r/PicsOfDeadKids or /r/spacedicks or /r/WhiteRights or /r/GreatApes or /r/IncestPorn (all NSFW/NSFL) do not, I can only conclude that there is something fundamentally threatening about the way in which SRS is inverting the rules of reddit's play space. It is especially ironic, however, given that so much discourse on the site relies on the carnivalesque and serves as an implicit critique of certain power structures, such as hegemonic American foreign policy or right-wing Tea Party activism or fundamentalist Christianity. But when those same tactics are used by some redditors to critique the carnival of reddit itself, part of the community closes ranks and defensively suggests that those people, those Others, are not a valid part of this space.

Where open-source idealism meets reality

Reddit is an open-source community. Anyone can use the platform to create his/her own subreddit or download the entire codebase from GitHub. And the creators of reddit embody much of the open-source ethos—for example, being extremely loathe to intervene in conflicts that occur on reddit or only reluctantly banning subreddits even if the content is objectionable. As mentioned above, many within the community view free speech as a sacrosanct, immutable right. Within the default subreddits in particular, there seems to be a distinct lack of understanding of the reason for First Amendment protection within the US—that it is not license to say anything without consequence on a privately owned and operated platform but that it specifically protects individuals from infringement from governmental institutions. I would argue that free speech is a particularly salient issue for the reddit community as an offshoot of its technolibertarian leanings and open-source ethos. But it is complicated by the algorithmic logic of the reddit platform, where the "hive-mind" can quickly take over any popular thread. Given that most redditors seem to disregard most of the upvoting/downvoting guidelines laid out in the site's reddiquette (instead upvoting things they agree with and downvoting things they do not, a point I discuss in detail in Chapter 4), threads can become quickly dominated by a limited set of responses, voices, and opinions—such is, perhaps, the inevitable consequence of any platform that uses "popularity" as a metric for making certain things more visible than others (see, for example, how the Google algorithm tends to flatten discourse around politics in Hindman, 2008). But for a community that views itself as prizing rational discourse over opinion-based reasoning, there's an awful lot of homogeneity in many reddit discussions if one only uses the default sorting mechanism ("best") to view threads.

The open-source community, broadly speaking, projects an egalitarian, non-hierarchical, meritocratic ethos (Bonaccorsi & Rossi, 2003). Within this space, individual contributions are subsumed by larger community concerns; knowledge, tools, and software are shared freely between participants. Open-source culture has its roots in early hacker culture of the 1970s and 1980s, chronicled in Stephen Levy's (2010) *Hackers* and more recently in Fred Turner's (2006) and Gabriella Coleman's (2013) work. At the core of much of the hacker ethos is a desire for information to be shared freely but that governments should not be able to trace individual, anonymous speech except in very limited and extreme cases. Hence, there is the open-source and hacker community's support of both the WikiLeaks project, anti-SOPA/PIPA efforts, and support for individuals such as Aaron Swartz—reddit founder and later technology activist (MacFarquar, 2013). So the hacker/technolibertarian/open-source ethos relies on a dual understanding of freedom of information and anonymity, with the latter being critical to ensure the former.

Given its geek pedigree, perhaps it is not surprising that reddit's rules view the sharing of private information (doxxing) as one of the greatest sins, except when doxxing involves breaking news events (as in the Boston Marathon Bombing) or the sharing of images of women "borrowed" from other social networks (as are regularly posted to /r/girlsinyogapants, etc.) or the sharing of photos of people in public places (such as at Walmart or on a beach). None of these constitute doxxing for most redditors. But the unmasking of Violentacrez or the actions of SRS do. Thus, there is a great irony in the way in which some reddit moderators and users employ freedom of information and "freeze peach" to wrongly presume freedom from consequences. Unfortunately, those at the greatest risk of doxxing are those who are already marginalized in other ways. In her analysis of the BlueSky forum, Lori Kendall describes how it reified a particular notion of geek masculinity. While geek masculinity challenged certain aspects of traditional hegemonic masculinity, it still served to implicitly repress non-dominant groups. She argues:

> The cultural connections of BlueSky among work, masculinities, computer use, and sociability ensure a male-dominated atmosphere regardless of the number of women present. For the most part BlueSky participants ... conform to dominant masculinity standards. They relate to each other in ways that support heterosexual masculinity (although not all identify as heterosexual) and in the process continue to objectify women. This demonstrates that even as members of nondominant groups increase, their effect on existing social norms may be minimal. Without the constant visual reminders provided by physical copresence, the dominant group can ignore or forget the presence of members of other groups. Because of this, members of subordinated groups may more easily join interactions with the dominant group *as long as they conform to its norms*. (Kendall, 2002, p. 81, emphasis in original)

Kendall's findings mirror my own: despite the presence of women, people of color, the disabled, LGBTQ individuals, and older adults, reddit's culture still serves to implicitly silence those who do not fit the "typical redditor" profile mentioned in Chapter 3. Those who are not the white, cisgendered, heterosexual men whom we often think of as "geeks" in the narrowest of terms are viewed as outsiders, and like Kendall's BlueSky forum, they have limited impact on the larger space. Their critiques are often dismissed—they are essentially delegitimized as being not really "redditors," in the same way that some within the gaming community might dismiss the numerous women among their ranks as not really being "gamers." This fact undermines the rhetoric of both reddit administrators and redditors themselves that champions the virtues of anyone being able to create his/her own subreddit—and mold reddit to his/her own purposes. Similiarly, as "outsiders" or "non-redditors," they are afforded less consideration when it comes to the potential sharing of personal information. Instead of being protected from doxxing, these individuals are very often treated as though they "should have known better" than to share images or information about themselves online—even if said information was never intended for spaces outside their own social networks.

Notes

1. SXSWi is the South by Southwest Interactive conference that gathers technology professionals and academics yearly (http://sxsw.com/interactive).
2. http://www.youtube.com/watch?feature=player_embedded&v=Nozkilj7bhE#!
3. Schaff created Imgur, a free image-hosting site, to host the large amount of images that serve as the majority of content shared by reddit members.
4. I must note that the earliest uses of the male gaze in feminist film theory focused exclusively on how white men looked at white women. As bell hooks (hooks, 1992) notes, feminism and feminist film theory have had a troubling history with race whereby women of color (and their particular experiences) have been erased in favor of the concerns of white women. My discussion of the male gaze (and women's experiences on reddit more broadly) is not meant to erase the very important ways in which race, class, sexual orientation, (dis)ability, gender identification, age, and so on intersect to create very particular (and often quite troubling) experiences of reddit. This chapter is merely a starting point of a much larger conversation that we need to have about these issues.
5. By applying the concept of the "male gaze" to reddit, I do not mean to suggest that redditors are passive consumers of the site's content or that there aren't multiple "gazes." And I do not wish to suggest that women do not have agency when being seen by men.
6. NB: There is a female-oriented version of this subreddit featuring images of male celebrities, /r/LadyBoners. /r/gentlemanboners features almost double the number of subscribers

as LadyBoners (240,805 to 123,730) as of April 4, 2014, and is ranked the 77th most popular subreddit to LadyBoners's 169th (see http://redditmetrics.com/r/gentlemanboners and http://redditmetrics.com/r/LadyBoners for more details).

7. Tits or GTFO (Get the Fuck Out) originated on 4chan but quickly spread to other internet spaces (such as online gaming) perceived to be the "domain" of men (SteveTR, 2011).
8. An example is a posting of comedian Dave Chappelle dressed as a crack addict with the words "Y'all got more of them compliments?" written on it (ZakDougall, 2014).
9. Although it is worth noting that the subreddit's FAQ suggests that they are *not* anti-feminist, but rather "pro-equality" as, "… it's meaningless to ask whether the Men's Rights Movement is anti-feminist until an agreement is reached on what feminism is" ("faq—MensRights," 2014).
10. This perception that men are subjected to false rape reports regularly and that those doing the reporting are not punished for their transgressions has led to activist efforts from both /r/MensRights and 4chan. For example, in 2013, individuals from these groups spammed Occidental College's anonymous online sexual assault web forum with false reports in an effort to raise awareness about this perceived "problem" (Culp-Ressler, 2013).
11. It is impossible not to see why approaching romance as a "game" might appeal to many redditors, as many have an intense interest in gaming and desire "rules" around social interactions, perhaps because of their self-avowed social awkwardness. This was noticed as much in a thread on /r/OkCupid (a subreddit dedicated to the OkCupid dating site), where someone asked if "negging"—that is, insulting women regularly while trying to seduce them—actually worked. One redditor noted that "… PUAs make talking to women seem like a video game," to which another responded, "There's a reason it's called Game Theory" (bel-a-rusian, 2014).
12. The photographer who shot this image has also filed a cease-and-desist claim against *Know Your Meme*, but the image is still widely available on the internet and appears regularly on reddit without permission.
13. From *Comedy Central Presents: Donald Glover* ("Donald Glover—Crazy Stories," 2010).
14. This meme has been a popular myth since the internet's beginnings. For more, see Lindell, 2010.
15. The term "white knight" typically refers to men who defend the honor of women in online spaces, usually with the hope of sexual fulfilment from said women (Don, 2013b). Interestingly, it is a disparaging term used by both anti-feminists (TRPs and MRAs) and feminists/so-called social justice warriors (SJWs). For the former group, the idea that a man would shill for feminist causes is reprehensible; for the latter, the idea that a woman needs "saving" by a man is equally appalling.
16. Something Awful was started in 1999 by Richard "Lowtax" Kyanka and is a kind of forerunner of 4chan and reddit. It was a hub for comedy and geek culture and popularized many memes in the days before either site existed. Posters were required to pay a monthly fee to access the forums ("Something Awful," 2014).
17. It should be noted that the Canadian blogger was later doxxed and, like the other three women listed, regularly received death and rape threats (Lewis, 2012; Liebelson & Raja, 2013; Romano, 2013; Watson, 2012).

· 7 ·
CONCLUSION

I knew my time on reddit was coming to a close the week I had a comment gilded (finally), received a Dogecoin tip, and was banned from posting to a subreddit. The first two were genuine, unexpected accomplishments. My comment mentioned something about how I thought allies of marginalized communities could learn a lot if they simply sat back and listened for a while before jumping in with their own opinions and that most folks don't take kindly to the Kanye West "Imma let you finish, but first let me say how you're wrong" approach (a common tactic on reddit). Surprisingly, it was this comment that was considered gold worthy that cheered me, as at the time I was deep in the throes of exploring the more disturbing elements of reddit that I detailed in the last chapter. The second event that happened resulted in a posting I had made regarding Bitcoin—one of the common tropes around the cryptocurrency is that people are quick to say how pretty much any development is "good for Bitcoin"—a comment I jokingly echoed in a posting after someone asked, "But, is this good for Bitcoin?" and I responded, "Everything is good for Bitcoin." Lo and behold, I found myself being tipped +10 Dogecoin (an offshoot of Bitcoin), which amounted to about $0.00005 USD. The Dogecoin folks are known on reddit as being especially generous, playful, and random (see, for example, their sponsorship of NASCAR driver Josh Wise's Dogecoin car [Bianchi, 2014]), and it was a pleasant surprise.

It was the last item, being banned from posting to a subreddit, however, which stung. By this point in my study, I was mostly spending time on some of the fringy-er meta-subreddits, especially /r/SubredditDrama, to preserve my sanity after so much time on the site and because it was a lot of fun to examine reddit culture and, admittedly, poke fun at its more extreme aspects with others. Around this same time (May 2014), a new set of default subreddits was announced, which would radically shift the content of the home page for individuals not logged in or without reddit accounts (and which would substantially remake /r/all). The most controversial change to the default subreddits was the addition of /r/TwoXChromosomes (2X), a women's issues subreddit with a healthy subscriber base—around 250,000 individuals. I had visited 2X in the past. It was a mostly mainstream women's issues subreddit that was inoffensive and benign, and it seemed a fairly supportive community for the people who posted regularly. 2X was not filled with Tumblr-like "social justice warriors" (SJWs) that often drew the ire of reddit's /r/TumblrInAction subreddit. Most postings revolved around relationship issues, requests for support with regard to sexual assault, and a kind of "feminism lite" feel. It was purposefully far less confrontational than SRS's FemPire in both its scope and approach to women's issues and seemed for the most part to be a friendly, if not particularly activist-oriented, community. Rarely did a 2X posting even make it to /r/all prior to its status change on the site—in fact, I would be surprised if many redditors even knew that the subreddit existed until its status change.

The moment community administrator /u/cupcake1713 announced the revised list of default subreddits, reddit lit up with outrage. Suddenly 2X became a shill for SRS and SJWs and represented proof positive for reddit conspiracy theorists (mostly from the men's rights and Red Pill subreddits) that the administrators were actually radical feminist shills out to destroy freeze peach (free speech) for everyone. 2X became brigaded by people downvoting all posts indiscriminately and trolling of the subreddit grew exponentially. Rape apologists, misogynists, and just plain assholes flooded the subreddit. Moderators of 2X were faced with a subscriber base that became increasingly angry, as they were not informed or consulted of the change before it was announced to the entire reddit community. 2X fought back with the tools they had at their disposal—posting and upvoting the most grotesque female-oriented discussions they could. Soon the front page of 2X was filled with several postings about digestive changes wrought by menstruation (aka "period shits"), vaginal odor, and explicit descriptions of menstrual flow. While these topics were not unfamiliar territory for 2X, the number of them increased exponentially as a result of the defaulting. Several of these posts ended up hitting the front page of the site, which further annoyed elements of the reddit community—this time, it was those who did not want their front pages "cluttered with gross stuff" (despite the fact that much, much, much worse content related to excrement, for example, was regularly posted to /r/AskReddit and /r/WTF on a daily basis). These redditors would derail any thread in 2X to mention their displeasure at its status change and were met with the suggestion that they were welcome to unsubscribe. For many of them, this was not enough, and the back and forth between subscribers before the defaulting and newcomers went on for some time.

The 2X drama acted as a glorious torrent of buttered popcorn for /r/SubredditDrama (SRD).[1] It was unreal—daily recaps were posted for 20+ days, chronicling every moment of 2X's moment in the spotlight. The drama was in some ways completely predictable, given the vocal numbers of redditors who held somewhat less-than-progressive views about women and were upset to see their presence on the reddit home page (/r/all, or reddit.com if they were not logged in). Since SRD attracts a wide swath of individuals across the reddit spectrum, the tone of the conversations around the 2X drama and resulting subscriber meltdown ranged from glee (likely from some of the MRA folks who populated the subreddit) to real sorrow (from those who had been 2X subscribers before the defaulting). I contributed a couple of links to the subreddit and was summarily banned for breaking their "no drama" rule.[2] I felt a twinge of sadness. Although I had spent little time on 2X, I felt a bit of shame that I would not be able to participate in the community unless I logged into the site with an alt (alternative account) or the moderators lifted the ban on my primary account.

Later that same week, I was accused of being an SRS "bigot," which I found both hilarious and sad. I was actually hoping to be called an "SRS shill" or "SJW," which I suspect could still happen after this book is published. I guess commenting on a couple of postings in the SRS FemPire and posting to SRD makes you a "bigot" in some strange universe. In some ways, all of these events signaled my full membership in the community—and also made it okay for me to take my leave.

The future of reddit

In September 2014, *Wired* reported that reddit's involvement in the celebrity photo hacking scandal resulted in enough money being generated through

Reddit Gold to pay for a month of the site's server costs (Greenberg, 2014). Reddit coffers profited greatly by the spike of traffic to /r/TheFappening, which was allowed to link to the stolen pictures long after other sites banned them. As traffic to the subreddit was reaching its peak, CEO Yishan Wong wrote a tone-deaf and oddly titled blog posting, "Every Man Is Responsible for His Own Soul," describing why reddit would not ban objectionable subreddits unless required to do so by law. He expressed sympathy that the platform had been used to distribute the images but suggested there would be no change to the site's policies as a result, "… because we consider ourselves not just a company running a website where one can post links and discuss them, but the government of a new type of community. The role and responsibility of a government differs from that of a private corporation, in that it exercises restraint in the usage of its powers" (yishan, 2014a). Wong then argued that individuals should make their own determination of the kind of material they'd like to see promoted, because "virtuous behavior is only virtuous if it is not arrived at by compulsion. This is a central idea of the community we are trying to create" (yishan, 2014a). This being reddit, one of the top comments about Wong's posting was the suggestion that its title sounded like something Jaden Smith would tweet.[3]

My first response to this posting was to pull a full Captain Picard facepalm.[4] How, I wondered, could reddit's leadership be so clueless as to assert that what they were doing was akin to running a government? No wonder some redditors have such a confused sense of what "free speech" means—it obviously stems from the misguided ways in which the site's administrators discursively position reddit as a "government" (or community or platform or at least something more than just a company) and then choose to let their citizenry run completely wild.

After thinking about Wong's statements further (which were later qualified and expanded, albeit unsatisfactorily, in a posting by one of the site's technical leads), I felt a sense of intense resignation. We have made this. We have chosen to have corporations, particularly US corporations, take the lead in everything from the internet's backbone to the services we use to access the web to the platforms and algorithms we use daily to organize, personalize, and repackage our information. We have allowed the Googles, Microsofts, Twitters, Facebooks, and, yes, the reddits to create the walled gardens where we play and labor. I suppose it should not be surprising in a country in which the government considers corporations people (Winkler, 2014) that these same corporations might turn around and declare themselves governments. So now what?

CONCLUSION 163

Reddit is at a critical juncture. If the site's leadership continues to push responsibility to self-police objectionable content on to community members and moderators, then it will continue to propagate. As a result, administrators will face increasing scrutiny from the media and will likely face legal action—from distributing either libelous or illegal material. In an effort to contain this kind of objectionable material, two reddits will emerge: one composed of subreddits which are sanitized enough to attract significant advertising attention and a second whose postings never make it to the front page and which remains effectively invisible to the casual browser. This is already happening, as a change to reddit's codebase in July 2014 made it possible for moderators of particular subreddits to be able to exclude their subreddit's material from showing up on /r/all—making it effectively invisible to non-subscribers (alienth, 2014). The "dark net" of reddit will make the company profitable because it will be extremely popular, but it will likely be disavowed or even spun off from the rest of reddit. To mitigate some of the disaster that is the reddit dark net, their public relations efforts will go into overdrive highlighting the community's activist efforts. RedditGifts will expand and become a platform for crowdsourcing charitable works of particular interest to the reddit community. And it will do some good, but it will hardly make up for all of the harm that is done by tacitly allowing the platform to become a haven for toxicity. Recycled and predominately image-based content will dominate the site's subreddits at an even greater rate, and a substantial portion of the population will eventually migrate away to other platforms, either in response to the continual reposts or because the administrators refuse to acknowledge the problematic culture they encourage.

Alternatively, reddit administrators can actively disallow certain uses of reddit.com. An easy first step, one that would no doubt cause great distress among some within the reddit community, would be to ban subreddits or user accounts that exist solely to promote hate speech, graphic content, or otherwise offensive material. Additionally, creating a much more robust policy expanding what might constitute doxxing (so that pictures of women, for example, that have been taken without their consent or shared without permission would be disallowed) could encourage a more welcoming environment for many individuals. Unfortunately, both administrators and moderators would likely perceive such changes as onerous and difficult—like playing a never-ending game of "whack-a-mole" by continually having to monitor and ban problematic subreddits. And this will inevitably annoy a number of reddit diehards who will decry the elimination of subreddits such as /r/girlsinyogapants,

/r/GreatApes, /r/incest, and /r/CandidFashionPolice as further evidence of the moderators' "fascist" tendencies. However, such a move might also provide a "teachable moment" for reddit administrators, as they could clarify why such material creates a hostile environment for site visitors and members. It would also help for administrators and developers to reconsider the ways in which the site's algorithm and the limited set of moderation tools on offer are complicit in propagating this toxicity.

Given the reluctance of administrators to take a stand in any meaningful way against the toxic material that populates their site, I am afraid the first scenario is most likely. It is frustrating that reddit, like so many other online spaces, creates a discursive environment where bullying, hatred, and bigotry are tacitly accepted in the guise of "free speech." This is not simply about anonymity functioning as a shroud behind which people hide to commit despicable acts—if anything, sites like reddit should remind us of the reality that they simply may make more visible preexisting tendencies in culture generally. But to hold the rest of the community hostage to those who use the site to make and share misogynist, transphobic, abelist, homophobic, and racist content is frankly unconscionable.

Despite all of the popular press and academic discussion that highlight the liminal and fluid nature of the web, I don't think reddit administrators really understand, or they choose not to acknowledge this reality. Reddit exists in a constellation of other platforms and communities. Part of what makes it a complex and fascinating point of cultural creation is that it centralizes a disparate and diverse collection of content scattered throughout the web. And as a platform, its power is clearly in amplifying, redistributing, and archiving material that may otherwise go unnoticed. So, when celebrity nude photos are stolen and posted to 4chan, reddit (and sister site Imgur) becomes the hub for their distribution and discussion—in part because the platform is not ephemeral like 4chan or Twitter. The administrators' unwillingness to address the central role reddit has played in regard to TheFappening and other events is genuinely disheartening. More than that, their continued refrain, "It's all in the users' hands," is not only disingenuous, but also just plain wrong. The net was only momentarily in the hands of the public before corporations moved in to control every aspect of our experience with it—from the equipment we use to access our ISP to the sites we use to communicate with one another. It is entirely possible for reddit administrators to take a stand on the ways in which reddit.com is used while still allowing the platform itself to remain "content neutral" and open source. Is it really that difficult to tell the 4500 subscribers to /r/GreatApes to take their racist and hate-filled speech back to Stormfront.org?

Reddit's administrators and moderators are not the only ones to blame for the site's toxicity. Redditors themselves are also responsible. As mentioned at the beginning of Chapter 5, one of the things that drew me to the space and kept me coming back was the irreverent humor and quick wit that many community members demonstrate. When institutions and those in power are the target, reddit's carnival is lively and engaging. But too frequently, those on the receiving end are those who are already disenfranchised and marginalized. This breaks a fundamental rule of the carnivalesque and comedy more broadly—that it should always "punch up" instead of down (Lynn, 2014). Thus, established social structures, hierarchies, and mores should be at the receiving end of the joke, not those who are already at the bottom of the social structure. Complicating all of this is the fact that a number of redditors likely view their identity as "geeks" as making them marginal to dominant images of hegemonic masculinity. While this may make them marginalized in some ways, it does not erase other forms of privilege they may enjoy. They may not understand (or refuse to acknowledge) why others might be offended by their use of humor to make thinly veiled racist, sexist, or classist statements. When this is pointed out, they often react hostilely, dismissing such concerns as ones that would only be voiced by "social justice warriors" or "white knights" or "not *real* redditors." Compounded with reddit's tendency to be a personalized echo chamber, this means that many redditors will have few opportunities to bear witness to lived experiences of others who do not share the same background. It is a conundrum, given the intense interest and charitable capacity they demonstrate at other moments.

Revisiting the politics of platforms

In the previous chapters, I have demonstrated that reddit culture encourages altruism, playfulness, meta-conversations, and creativity among its members. All of this, I argue, has to do with the people inhabiting the space and the politics underlying reddit's platform. These good things are not negated by the flip side of what happens in this space—it is plagued by the same kinds of superficiality, problematic discourse, and cynicism that infect most online spaces today. So why does the reddit culture support, on the one hand, a kind of playfulness and altruistic behavior and on the other hand, a lot of alienating and discriminatory speech?

The Greater Internet Fuckwad Theory (GIFT) is a popular explanation of the reason that internet spaces seem to bring out the worst in all of us (Brad, 2013b). The *Penny Arcade* comic that describes GIFT postulates that the "normal" person combined with an audience and anonymity will inevitably turn into a "fuckwad" who spews racist and sexist garbage when given a chance. Redditors and others within online spaces often mention GIFT—and specifically anonymity—as the main cause for such offensive speech online.

However, this explanation is wanting. I would argue that GIFT places too much blame at the feet of anonymity. As Lisa Nakamura (2013) noted in her 2013 keynote for the Association of Internet Researchers conference, the GIFT formulation allows people to get off too easily. It is not just community spaces like reddit or comment sections of newspaper sites (Hughey & Daniels, 2013) or Xbox Live chatter (K. L. Gray, 2012) that drifts into fuckwad territory—it is culture more broadly. And even those spaces that require Facebook logins, for example are not exactly the paragons of deliberative discourse, nor do people refrain from being "fuckwads" just because their name is attached. One only needs to visit the online comment section of the *Chicago Tribune* to be disabused of that notion.

Additionally, dismissing those who rabble rouse in this way as just "normal people" hiding behind anonymity is too simplistic. Many of us, even when anonymous, would not spew hate-filled invective when given the chance. Unfortunately, these "normal people" reflect a very real and very disturbing trend, whereby those who have often been the most privileged to speak, act, and do are suddenly seeing themselves as oppressed by feminists, people of color, and other groups. For example, research suggests that white individuals are increasingly viewing themselves as routinely targeted by racism (Norton & Sommers, 2011). And as other scholars have argued, infrastructural power relationships are equally responsible for the ways in which online communities may become spaces for oppression:

> It is tempting to suggest that the networked mass sites that allow the uncontrolled flow of communication between so many people are responsible, at least in part, for the unleashing of a ramped-up form of sexism via a new mob mentality. But the simplifications of technological determinism always miss the social power relations that are prior to and circulating through multiple cultural institutions. (Humphreys & Vered, 2014, p. 4)

So, if it is not strictly anonymity, perhaps speech like this is the direct result of its public nature. It is true that in terms of reddit, the karma mechanism encourages a potential bandwagon effect, whereby the top-rated content and comments achieve more fame and become more visible because they are top rated. The nature of "drama" on the site (and the internet more broadly) is tied inextricably to the audience for whom that drama plays out. And in some cases, the spectators revel in the drama, pulling out the popcorn (or the folding chair) to watch a good show. But, again, I do not think this entirely explains the phenomenon.

What is missing from GIFT is the role that platform politics plays. How do platforms create technological logics that encourage certain kinds of discourse and silence others? Anyone can use the reddit platform to create her own subreddit or even her own version of reddit by downloading the platform code from GitHub. And certainly the creators of reddit embody much of the open-source ethos—for example, being extremely loathe to intervene in conflicts that occur on reddit in the name of free speech and only reluctantly banning subreddits even if the content is objectionable. But the algorithmic logic of reddit—where upvotes and downvotes determine visibility on the site—also has a huge impact.

The end of participatory culture?

Reddit's problems reflect a larger cultural crisis. When Henry Jenkins first unveiled the concept of participatory culture in 2006, Web 2.0 was still in its infancy (Jenkins, Purushotma, Weigel, Clinton, & Robison, 2009). Twitter had just been founded, Facebook was only two years old, and YouTube had gone online only the year before. Blogging platform Wordpress was a mere three years old. Instagram, Pinterest, Snapchat, and Tumblr were not around yet. Reddit was only a year old. Social networking sites were relatively new, and the creation and sharing of user-generated content was mostly done through blogs, newsgroups, online bulletin boards, and the like. Still, the increasing availability and decreasing cost of production tools as well as the internet's vast distribution networks were blurring the distinction between media consumer and media producer. And so Jenkins's emphasis on the "participatory" nature of media seemed important and valuable, as it reflected a real shift in the ways in which we engaged with media content and each other. But things have changed.

While I was finishing this book, news of the "Facebook Contagion" study had just been released, igniting a firestorm of questions regarding research ethics, corporate responsibility, and algorithmic politics among academics and the general public (Meyer, 2014). Perhaps the most searing critiques of the research project came from digital humanist Nick Montfort in a blog posting written as a series of questions, ending with "Why do we think that we can fix Facebook? Why did people who communicate and learn together, people who had the world, leave it, en masse, for a shopping mall?" (Montfort, 2014). And this is the crux of where we are at this moment in time in regard to participatory culture (and mediated culture generally). It has been commodified and resold to those of us who make, share, and communicate in this space. I wonder whether this means that the end of "participatory culture" draws near. It is not that individuals are no longer participating or creating or sharing; it is that we have come to understand that such actions do not necessarily encourage greater engagement with the world at large or are inherently more democratic or ensure a more peaceful and just future. If anything, "participatory culture" has enabled some of the more disturbing elements of human nature to become more visible and perhaps gain more traction than they might have otherwise. Fortunately, such visibility also makes it possible to fight back against these tendencies, but it is a fight.

All of this may sound grim. However, I do not want to understate the importance that reddit's platform, in particular, has played and could continue to play in repudiating the real-name/one-identity policies of many social networking sites. Such policies, I believe, are fundamentally antiproductive and antithetical to the web's potential for identity play and meaningful community discourse. Pseudoanonymity encourages redditors to build online reputations that express humor, irreverence, skepticism, and authenticity. It may also embolden individuals to provide and offer social support and engage in play with less concern that these activities will be tied to a singular, "one-name" identity. In this sense, reddit represents a throwback to the earlier days of the web, when Howard Rheingold's (2000) virtual community and Sherry Turkle's (1995) identity play were the norm. But it also makes individuals accountable to the community in ways that complete anonymity does not. I would not be so bold as to suggest that reddit fixes all of the problems people might have with contemporary online culture, nor could I in good faith suggest that it offers a true alternative (as it is still owned and operated by a large corporation and dependent on the infrastructure of the internet with all of its attendant difficulties). I do think it might offer a glimmer of hope—but only if its administrators and moderators take concrete steps to mitigate the culture's tendency toward bigotry, misogyny, racism, and so on.

Future directions

As I stated in the introduction, this book's goals were twofold. I wished to problematize some of the ways in which we thought about participatory culture, and I suggested that documenting spaces like reddit contemporaneously could provide critical information about how members create and make meaning of an online culture. I also argued for the importance of considering how the relationship between people and technology is productive and co-constitutive. By employing ethnographic methods, I hoped to create a rich, multifaceted portrait of culture on reddit—and the ways in which this culture is contested by its members as well as shaped by the politics of its platform.

At a conference at which I presented an early version of the previous chapter, I briefly mentioned some of the feedback I had received from moderators regarding the paucity of tools for ensuring quality discussions on reddit. After the presentation, I was asked what kinds of tools or changes to the reddit algorithm could help this problem. I responded with an honest answer, "That's a great question, but I don't know." Like many things reddit-related, this project has raised more questions for me than it has answered.

There are many, many different directions that future research into spaces like reddit needs to go. In terms of reddit, specifically, the cultures of individual subreddits are deserving of exploration. As I mentioned in the introduction, reddit is both *a* culture and *many* cultures, and the multiple, diverse subcultures that subreddits represent are worthy of their own detailed examination. The nature and function of humor and play on reddit also deserve further attention. Broadening this out, it would be useful for us to have a better understanding of the ways in which play is enacted and how it can be encouraged and supported on other platforms.

In terms of the less savory aspects of participatory culture, the question of why geek culture (and reddit specifically) seems to serve as an important focal point for misogyny and many anti-feminist men's rights groups is deserving of further study. Another critical concern is an examination of the ways that hate groups such as Stormfront have enlisted the language of memes and play as a recruitment tool on these sorts of spaces (and done so successfully)—and how communities can prevent this from happening. There is also a larger question of community responsibility and governance with regard to participatory platforms. Should reddit and other online spaces continue community "policing" of objectionable content? Or is this simply a clever way of encouraging community "ownership" while actually avoiding administrative accountability?

It is critical that as scholars we continue to move beyond the "hype" that so often characterizes media discourse and our own first encounters with new technologies. Thus, encouraging deep and long-term engagement with online environments and moving beyond explanations that reify a kind of "digital dualism" are important. So too is promoting the continued exploration of how individuals, communities, and technologies are co-constitutive. Only then can we adequately convey the richness of these spaces and preserve them in some fashion for future generations.

Notes

1. SRD engages in mirthful postings about any drama in other subreddits. And, as discussed in Chapter 4, while the unstated idea around SRD was merely to comment on said drama instead of having the drama occur in the subreddit itself, drama did populate the subreddit with some frequency; hence the meta-meta subreddits discussing said drama, /r/SubredditDramaDrama.
2. Apparently, 2X (like some other subreddits) does not like any links to it from inside reddit—especially those that come from meta-subreddits—even if no vote brigading is occurring.
3. Will Smith's son Jaden is known for his perplexing Twitter feed, which resembles what might happen if a stoner started spouting Zen koans, but instead of discussing the meaning of life, he discusses Shia LeBeouf, trees, and song lyrics in oddly capitalized tweets.
4. http://knowyourmeme.com/memes/facepalm

GLOSSARY

Admins/Administrators: reddit employees who have ultimate control over the site's content and user base; responsible for enforcing site-wide rules and policy changes

AFAIK: "As far as I know"; used to suggest the comments that follow are truthful only insofar as the redditor him/herself has verified

AMA: "Ask me anything"; also the name of a popular subreddit and the events it hosts in which redditors can ask a question of a popular celebrity; often used in conjunction with "IAMA"

AMAA: "Ask me almost anything"; suggests a willingness to answer almost any question on a topic

CMV: "Change my view" is used to indicate that a redditor is interested in engaging in a debate around a belief or position he/she holds

Copypasta: copy-and-pasted text that is shared because of its verbosity or linguistic uniqueness or perceived ridiculousness given its original context

DAE: "Does anybody else" or "Does anyone else"; often used to ask others about unique experiences or habits that they think may be unique

Doxxing: sharing private information about another individual

ELI5: "Explain like I am 5 years old"; a request to explain a complex concept or situation in the simplest terms possible

FFS: "For fucks' sake"; used to indicate disgust

Friendzone: a friendship in which one person desires a sexual or romantic relationship but the other one does not

FTFY: "Fixed that for you"; used to indicate a humorous "fix" to an image or comment made by the OP
GGG: Good Guy Greg or Good Girl Gina; two popular image macros used to indicate the actions of a good person
IAMA: "I am a ..."; often used in connection with AMAs
IANAD: "I am not a doctor"; often used to suggest that the medical advice about to be given is no substitution for consulting a physician
IANAL: "I am not a lawyer"; often used to suggest that the legal advice about to be given is no substitution for consulting a lawyer
IIRC: "If I recall correctly"; used to provide information about the topic at hand; similar to AFAIK
ITT: "In this thread"; usually used in a pejorative manner to describe what the commenter believes is the general opinion or tenor of the discussion at hand
Karma or karma points: the total number of upvotes (minus downvotes) a given redditor's contribution has received
Karmawhoring: the perceived act of reposting or sharing popular content just for karma
Mods/Moderators: community members who volunteer to run a subreddit, enforcing its rules
MRA: Men's Rights Activist; used to indicate that someone is a member of the /r/MensRights subreddit or anyone who might espouse the same sorts of beliefs
MRW/MFW/HIFW: "My reaction when ...," "My face when ...," and "How I feel/felt when ...," respectively; usually used in combination with a reaction GIF
Narwhal bacons at midnight: originally offered as a kind of "secret passphrase" that one could use to determine whether someone else was a redditor; now regarded mostly as a pejorative term to describe those who are overly invested in reddit as an identity
NSFL: "Not safe for life"; indicates that the posting or subreddit offers gory or gross content
NSFW: "Not safe for work"; indicates that the posting or subreddit offers sexual or otherwise offensive content
OC: "Original content"; used to indicate that the material is not a repost and was created in some way by the OP
OP: "Original poster"; used to indicate the person who starts a thread
PUA: "Pickup artist"; used to suggest someone who practices seducing other individuals, often using techniques from author Neil Strauss's book *The Game*
RedPiller: person who is a member of the /r/TheRedPill subreddit
RES: "Reddit Enhancement Suite" a browser add-on that significantly improves reddit's user interface by adding a number of interface enhancements that allow a user to tag others and customize the subreddits that appear on /r/all
Self-post/text-post: a posting that consists of just text or text and links for which the redditor receives no karma points
[Serious]: indicates that the OP wants only serious replies to his/her posting; commonly used in /r/AskReddit to discourage individuals from answering humorously
Shadowban: a ban that can be issued only by the site's administrators in which the person being banned is unaware of the ban; often used to prevent spammers from creating another account because it looks as though they can submit content as usual

SJW: "Social justice warrior"; used disparagingly to describe someone who is perceived as being too "politically correct" or interested in social justice issues

SRS: "Shit Reddit Says"; a subreddit dedicated to making offensive content shared by redditors more visible; pejoratively used to suggest someone is an SJW or being "politically correct"

Sub/subreddit: an individual reddit community organized around a topic of interest

TIFU: "Today I fucked up"; also a popular subreddit in which people share stories about how they screwed up

TIL: "Today I learned …"; used to indicate something that the OP learned (also the name of a popular subreddit in which such information is shared).

TL;DR: "Too long; didn't read"; either used to indicate that the commenter did not read the original posting as it was too long or used by the OP to offer a shorter summary of a lengthier posting

White knight: disparaging term used to suggest someone is espousing "politically correct" beliefs simply to curry favor with potential sexual partners

YSK: "You should know …"; used to indicate a fact or life skill that others feel is worth sharing

REFERENCES

1to34. (2013, May 26). In response to the article, "The dark side of Reddit's Gonewild." Retrieved April 5, 2014, from http://www.reddit.com/r/AdviceAnimals/comments/1f31ns/in_response_to_the_article_the_dark_side_of/

4chan frequently asked questions. (2014). Retrieved July 6, 2014, from http://www.4chan.org/faq

4chan rules. (2014). Retrieved July 6, 2014, from http://www.4chan.org/rules

_black. (2007, November 26). Greenpeace are having a vote to name a whale they have "adopted." All the options are the names of ancient gods of the sea. And then there's "Mister Splashy Pants." Please vote "Mister Splashy Pants." Retrieved March 10, 2014, from http://www.reddit.com/comments/61gqb/greenpeace_are_having_a_vote_to_name_a_whale_they

_JaredLeto. (2013, October 25). Jared Leto AMA. Retrieved April 20, 2014, from http://www.reddit.com/r/IAmA/comments/1p7vqv/jared_leto_ama/

Aarseth, E. (2007). *I fought the law: Transgressive play and the implied player.* Paper presented at Situated Play: the Third Annual Conference of the Digital Games Research Association (DiGRA), Tokyo, Japan.

Abad-Santos, A. (2013, April 22). Reddit's "Find Boston Bombers" founder says "it was a disaster" but "incredible." Retrieved March 10, 2014, from http://www.thewire.com/national/2013/04/reddit-find-boston-bombers-founder-interview/64455/

Abbate, J. (2012). *Recoding gender: Women's changing participation in computing.* Cambridge, MA: The MIT Press.

About bit of news. (2014). Retrieved May 19, 2014, from http://bitofnews.com/about/
About reddit. (2014, July 14). Retrieved July 16, 2014, from http://www.reddit.com/about/
Adam, A. (2002). Cyberstalking and internet pornography: Gender and the gaze. *Ethics and Information Technology, 4*(2), 133–142. doi: 10.1023/A:1019967504762
Addyct. (2014, June 22). Meta: What motivates it? [Comment]. Retrieved July 1, 2014, from http://www.reddit.com/r/TheoryOfReddit/comments/28rh9m/meta_what_motivates_it/cidq94p
Al_Simmons. (2013, April 14). Guys, please don't go as low as this. Retrieved April 20, 2014, from http://www.reddit.com/r/cringe/comments/1cbhri/guys_please_dont_go_as_low_as_this/
alanwins. (2014, June 23). Boy meets fire [Comment]. Retrieved July 20, 2014, from http://www.reddit.com/r/funny/comments/28uomp/boy_meets_fire/cieoow3
Alexander, J. C. (2004). From the depths of despair: Performance, counterperformance, and "September 11." *Sociological Theory, 22*(1), 88–105. doi: 10.1111/j.1467-9558.2004.00205.x
Alfonso, F., III. (2014, February 7). Creepshots never went away—we just stopped talking about them. *The Daily Dot*. Retrieved April 5, 2014, from http://www.dailydot.com/lifestyle/reddit-creepshots-candidfashionpolice-photos/
alienth. (2013, November 19). Glorious masterrace hear me [Comment]. Retrieved April 25, 2014, from http://www.reddit.com/r/gloriouspcmasterrace/comments/1r01ny/glorious_masterrace_hear_me/cdi9ld6
———. (2014, July 7). Experimental reddit change: subreddits may now opt-out of /r/all. Retrieved September 12, 2014, from http://www.reddit.com/r/changelog/comments/2a32sq/experimental_reddit_change_subreddits_may_now/
AlleaGirl. (2014, May 3). "This bitch is such an attention whore" /Gonewild [Comment]. Retrieved May 3, 2014, from http://www.reddit.com/r/SubredditDrama/comments/24l9ny/this_bitch_is_such_an_attention_whore_gonewild/ch8fi3h
Althoff, T., Danescu-Niculescu-Mizil, C., & Jurafsky, D. (2014). *How to ask for a favor: A case study on the success of altruistic requests.* Paper presented at the 8th International Conference on Weblogs and Social Media (ICWSM-14), Ann Arbor, MI. Retrieved from http://arxiv.org/abs/1405.3282
amlaviol. (2014, April 17). In response to all the higher education haters … (FIXED). Retrieved April 18, 2014, from http://www.reddit.com/r/AdviceAnimals/comments/239py7/in_response_to_all_the_higher_education_haters/
Andrejevic, M. (2008). Watching television without pity: The productivity of online fans. *Television & New Media, 9*(24), 24–46.
anniebeeknits. (2014, May 5). Some days diabetes feels as much like a mental illness as a physical one. Retrieved May 15, 2014, from http://www.reddit.com/r/diabetes/comments/24t8ir/some_days_diabetes_feels_as_much_like_a_mental/
antizeitgeist. (2012, October 2). Comment on posting: Girl walks off stage (super cringe after 2 minutes). Retrieved February 20, 2014, from http://www.reddit.com/r/cringe/comments/10uyi6/girl_walks_off_stage_super_cringe_after_2_minutes/c6gzx22
antsav888. (2011, December 25). Why is reddit's search engine so terribly bad? Retrieved July 17, 2014, from http://www.reddit.com/r/askreddit/comments/nqxv8/

AquilaGlobumAncora. (2013, July 8). College Liberal—Heard this today. Retrieved April 11, 2014, from http://www.reddit.com/r/AdviceAnimals/comments/1hvyk5/college_liberal_heard_this_today/

Arbitrary day 2014. (2014). Retrieved May 10, 2014, from http://redditgifts.com/exchanges/arbitrary-day-2014/

ArchangelleDworkin. (2012, April 12). [META] This world has some twisted fucks in it. Retrieved April 30, 2014, from http://www.reddit.com/r/ShitRedditSays/comments/s73yu/meta_this_world_has_some_twisted_fucks_in_it/

ArchangelleSamaelle. (2013, November). [META] SRS FAQ. Retrieved March 10, 2014, from http://www.reddit.com/r/ShitRedditSays/comments/o0pdv/meta_srs_faq/

ArchangelleStrudelle. (2012, August 11). [META] In celebration of the 1-year anniversary of SRS-as-we-know-it, here is A Not-That-Brief History of SRS. Retrieved May 5, 2014, from http://www.reddit.com/r/ShitRedditSays/comments/y2wag/meta_in_celebration_of_the_1year_anniversary_of/

askusmar. (2011, December 1). My Secret Santa is sending me on A CRUISE! Retrieved May 10, 2014, from http://redditgifts.com/gallery/gift/my-secret-santa-sending-me-cruise/

Attwood, F. (2011). Through the looking glass? Sexual agency and subjectification in cyberspace. In C. S. R. Gill (Ed.), *New femininities? Postfeminism, neoliberalism and identity* (pp. 203–214). London: Palgrave.

aux. (2012). Good Guy Greg. *Know Your Meme*. Retrieved April 5, 2014, from http://knowyourmeme.com/memes/good-guy-greg

AyChihuahua. (2012, May 28). So I was at a baby shower yesterday, and these were in some of the ice cubes ... Apparently I was the only one that thought it was hilarious [Comment]. Retrieved July 5, 2014, from http://www.reddit.com/r/funny/comments/u9ke5/so_i_was_at_a_baby_shower_yesterday_and_these/c4tinhy

Bakhtin, M. (1984). *Rabelais and his world* (H. é. Iswolsky, Trans.). Bloomington: Indiana University Press.

Banks, D. A. (2013, December 20). Very serious populists. *The New Inquiry*. Retrieved July 15, 2014, from http://thenewinquiry.com/essays/very-serious-populists/

Barbrook, R. (1998). The hi-tech gift economy. *First Monday*, *3*(12). http://firstmonday.org/ojs/index.php/fm/article/view/631/552 doi:10.5210/fm.v3i12.631

Barthes, R. (1977). *Elements of semiology* (A. Lavers & C. Smith, Trans.). New York: Hill & Wang.

Baum, K., Catalano, S., Rand, M., & Rose, K. (2009, January). *Stalking victimization in the United States* (NCJ 224527). Washington, DC: U.S. Department of Justice. Retrieved from http://www.ovw.usdoj.gov/docs/stalking-victimization.pdf

Baym, N. K. (2000). *Tune in, log on: Soaps, fandom, and online community*. Thousand Oaks, CA: Sage.

BearkeKhan. (2014, February 1). GoneWildResearcher new bot on the block. Retrieved April 5, 2014, from http://www.reddit.com/r/botwatch/comments/1wq2vu/gonewildresearcher_new_bot_on_the_block/

Beaulieu, A. (2004). Mediating ethnography: Objectivity and the making of ethnographies of the internet. *Social Epistemology*, *18*(2–3), 139–163.

bel-a-rusian. (2014, May 19). Guy admits he's negging me in first paragraph. Proceeds to neg throughout message. "Swears" he's not a douche. Retrieved May 19, 2014, from http://www.reddit.com/r/OkCupid/comments/25wl9a/guy_admits_hes_negging_me_in_first_paragraph/chlfh5v

Believeinfacts. (2009, December 14). My package required surgery! Retrieved May 5, 2014, from http://redditgifts.com/gallery/gift/my-package-required-surgery/

Bennett, W. L. (2004). Gatekeeping and press-government relations: A multigated model of news construction. In L. L. Kaid (Ed.), *Handbook of political communication research* (pp. 283–314). Mahwah, NJ: Lawrence Erlbaum Associates.

bennyschup. (2014, May 18). What is something that screams douchebag from a person's appearance? Retrieved May 21, 2014, from http://www.reddit.com/r/AskReddit/comments/25v56j/what_is_something_that_screams_douchebag_from_a/

berbertron. (2011, August 16). 2 AM ICE CHILI SHOWER. Retrieved August 1, 2013, from http://www.reddit.com/r/pics/comments/jlbdf/2_am_ice_chili_shower/

Bergstrom, K. (2011). "Don't feed the troll": Shutting down debate about community expectations on Reddit.com. *First Monday*, 16(8).

Bernstein, M. S., Monroy-Hernández, A., Harry, D., André, P., Panovich, K., & Vargas, G. (2011). 4chan and/b: An analysis of anonymity and ephemerality in a large online community. *Proc. ICWSM2011*, 50–57.

Bianchi, J. (2014, May 16). Reddit users help get Dogecoin car and Josh Wise into NASCAR All-Star Race. *SB Nation*. Retrieved July 19, 2014, from http://www.sbnation.com/nascar/2014/5/16/5725312/reddit-dogecoin-josh-wise-danica-patrick-fan-vote-2014-nascar-all-star-race

Bidgood, J. (2013, April 25). Body of missing student at Brown is discovered. *The New York Times*. Retrieved March 10, 2014, from http://www.nytimes.com/2013/04/26/us/sunil-tripathi-student-at-brown-is-found-dead.html

Big Bang Theory Irony. (2012, February 25). Retrieved May 21, 2014, from http://www.reddit.com/r/AdviceAnimals/comments/q5rsc/big_bang_theory_irony/

Bijker, W. E., Hughes, T. P., & Pinch, T. (Eds.). (2012). *The social construction of technological systems: New directions in the sociology and history of technology* (Anniversary ed.). Cambridge, MA: The MIT Press.

Bivens, K. (2008). The internet, mobile phones and blogging: How new media are transforming traditional journalism. *Journalism Practice*, 2(1), 113–129.

BlackbeltJones. (2012, September 7). Since the frontpage has turned to shit, is there a faster way to shit up the comment threads? YES THERE IS!! Retrieved May 26, 2014, from http://www.reddit.com/r/circlebroke/comments/zivl4/since_the_frontpage_has_turned_to_shit_is_there_a/

Bobby_Ooo. (2012, December 27). I told my girlfriend that I didn't want to have sex because I was too tired. So right before we got it on she said this. Retrieved April 4, 2014, from http://www.reddit.com/r/AdviceAnimals/comments/15jsan/i_told_my_girlfriend_that_i_didnt_want_to_have/

Boellstorff, T., Nardi, B., Pearce, C., & Taylor, T. L. (2012). *Ethnography and virtual worlds: A handbook of method*. Princeton, NJ: Princeton University Press.

Bogers, T., & Wernersen, R. N. (2014). How "social" are social news sites? Exploring the motivations for using reddit.com. Paper presented at the iConference, Berlin, Germany. Retrieved from http://toinebogers.com/content/publications/bogers-iconf2014-social-news-motivation.pdf

Bonaccorsi, A., & Rossi, C. (2003). Why open source software can succeed. *Research Policy, 32*, 1243–1258. doi: 10.1016/S0048-7333(03)00051-9

Borsook, P. (2001). *Cyberselfish: A critical romp through the terribly libertarian culture of high tech.* New York: PublicAffairs.

Bourdieu, P. (1977). *Outline of a theory of practice.* Cambridge: Cambridge University Press.

Brad. (2012a). Ridiculously Photogenic Guy / Zeddie Little. *Know Your Meme.* Retrieved July 15, 2014, from http://knowyourmeme.com/memes/ridiculously-photogenic-guy-zeddie-little

———. (2012b). Scumbag Girl / Scumbag Stacy. *Know Your Meme.* Retrieved June 30, 2014, from http://knowyourmeme.com/memes/scumbag-girl-scumbag-stacy

———. (2013a). College Liberal. *Know Your Meme.* Retrieved April 11, 2014, from http://knowyourmeme.com/memes/college-liberal

———. (2013b). Greater Internet Fuckwad Theory. *Know Your Meme.* Retrieved July 15, 2014, from http://knowyourmeme.com/memes/greater-internet-fuckwad-theory

———. (2013c). Horse-Sized Duck. *Know Your Meme.* Retrieved April 18, 2014, from http://knowyourmeme.com/memes/horse-sized-duck

———. (2013d). Just Go On The Internet and Tell Lies. *Know Your Meme.* Retrieved June 26, 2014, from http://knowyourmeme.com/memes/just-go-on-the-internet-and-tell-lies

———. (2013e). Overly Attached Girlfriend. *Know Your Meme.* Retrieved April 5, 2014, from http://knowyourmeme.com/memes/overly-attached-girlfriend

Braitch, J. (2010). The digital touch: Craft-work as immaterial labour and ontological accumulation. *Ephemera: Theory and Politics in Organizations, 10*(3/4). http://www.ephemerajournal.org/contribution/digital-touch-craft-work-immaterial-labour-and-ontological-accumulation

Brandtlyc. (2014, 20 February). AMBER ALERT PARK RIDGE: r/chicago can help! Retrieved February 20, 2014, from http://www.reddit.com/r/chicago/comments/1ygvuu/amber_alert_park_ridge_rchicago_can_help/

BraveSirRobin. (2011). OP is a Faggot. *Know Your Meme.* Retrieved January 13, 2014, from http://knowyourmeme.com/memes/op-is-a-faggot

BronPinchot. (2012, March 29). I am Bronson Pinchot from Historic Architecture—"The Bronson Pinchot Project on DIY"—to the most obscure Balki trivia—to anything else from the Langoliers to what have you. But please no questions about Rampart. (That is where I draw the line). Retrieved April 20, 2014, from http://www.reddit.com/r/IAmA/comments/rk60c/i_am_bronson_pinchot_from_historic_architecture/

BronzeLeague. (2013, March 28). RES, voting, and reaction GIF's. Retrieved May 30, 2014, from http://www.reddit.com/r/circlebroke/comments/1b7aux/res_voting_and_reaction_gifs/

Brown, K. (2012). *Everyday I'm Tumblin': Performing online identity through reaction GIFs* (Unpublished master's thesis). The School of the Art Institute of Chicago, IL. Retrieved from http://www.academia.edu/1975475/Everyday_Im_Tumblin_Performing_Online_Identity_through_Reaction_GIFs

Bruns, A., & Highfield, T. (2012). Blogs, Twitter, and breaking news: The produsage of citizen journalism. In R. A. Lind (Ed.), *Produsing theory in a digital world: The intersection of audiences and production in contemporary theory* (pp. 15–32). New York: Peter Lang.

Buchanan, R. (1998). Wicked problems in design thinking. In V. Margolin & R. Buchanan (Eds.), *The idea of design* (pp. 3–20). Cambridge, MA: The MIT Press.

Bucher, T. (2014). About a bot: Hoax, fake, performance art. *M/C Journal, 17*(3).

buckysbitch. (2013, March 12). I just adopted an abandoned cat. This is the first thing he did when we got home and I let him out of his crate. Retrieved August 11, 2013, from http://www.reddit.com/r/aww/comments/1a7awt/i_just_adopted_an_abandoned_cat_this_is_the_first/

Burrill, D. A. (2008). *Die tryin': Videogames, masculinity, culture.* New York: Peter Lang.

Caillois, R. (2001). *Man, play and games* (M. Barash, Trans.). Urbana: University of Illinois Press.

Callon, M., & Latour, B. (1981). Unscrewing the big Leviathan: How actors macro-structure reality and how sociologists help them do it. In K. D. Knorr Cetina & A. Cicourel (Eds.), *Advances in social theory and methodology: Towards an integration of micro and macro sociologies* (pp. 276–303). London: Routledge.

calmbatman. (2013, July 13). What is the "Digg Migration"? Retrieved June 2, 2014, from http://www.reddit.com/r/OutOfTheLoop/comments/1i8mp7/what_is_the_digg_migration/

CaptionBot. (2013, September 4). you all fell for it ... suckers. Retrieved November 1, 2013, from http://www.reddit.com/r/AdviceAnimals/comments/1lq14i/you_all_fell_for_it_suckers/

Carmichael, R. (2013, April 3). Darien Long is gone but not forgotten at Metro Mall. *Creative Loafing*. Retrieved May 14, 2014, from http://clatl.com/atlanta/darien-long-is-gone-but-not-forgotten-at-metro-mall/Content?oid=7935813

CationBot. (2013, August 1). CationBot comments on I also deliver pizzas. Retrieved August 2, 2013, from http://www.reddit.com/r/AdviceAnimals/comments/1jl1nl/i_also_deliver_pizzas/cbfqrf3

Chen, A. (2012, October 12). Unmasking reddit's Violentacrez, the biggest troll on the web. Retrieved August 1, 2013, from http://gawker.com/5950981/unmasking-reddits-violentacrez-the-biggest-troll-on-the-web

Christensen, H. S. (2011, February 7). Political activities on the Internet: Slacktivism or political participation by other means? *First Monday*. Retrieved June 20, 2014, from http://firstmonday.org/ojs/index.php/fm/article/view/3336/2767

Churba. (2014, May 2). 30 Rock had a one off character, Jerem, that reminds me of the people I have interacted with on reddit—[0:28] [Comment]. Retrieved May 2, 2014, from http://www.reddit.com/r/videos/comments/24iowz/30_rock_had_a_one_off_character_jerem_that/ch7sfsn

CIRCLEJERK_BOT. (2012, September 22). Post responses here. Retrieved August 1, 2013, from http://www.reddit.com/r/circlejerk_bot/comments/10aoct/post_responses_here/

Clifford, J., & Marcus, G. E. (1986). *Writing culture: The poetics and politics of ethnography.* Berkeley: University of California Press.

Coleman, E. G. (2013). *Coding freedom: The ethics and aesthetics of hacking.* Princeton, NJ: Princeton University Press.

colocito. (2014, February 2). A reminder about Reddit-wide rules [Comment]. Retrieved April 20, 2014, from http://www.reddit.com/r/AdviceAnimals/comments/1wui8n/a_reminder_about_redditwide_rules/cf5hs6r

ColtenW. (2012). Sweet Brown / Ain't Nobody Got Time for That. *Know Your Meme.* Retrieved May 21, 2014, from http://knowyourmeme.com/memes/sweet-brown-aint-nobody-got-time-for-that

Comes to reddit Only to bitch about it. (n.d.). Retrieved June 24, 2014, from http://www.quickmeme.com/meme/35sfyo

concerneddad1965. (2012, September 14). I am the father/Redditor who lost his family after it came to light that my son was sexually abusing our dog, Colby. I have some good news for everyone: COLBY IS SAFE. But there is still the question of what to do with my son? Retrieved March 10, 2014, from http://www.reddit.com/r/Askreddit/comments/zw3j9/i_am_the_fatherredditor_who_lost_his_family_after/

Connell, R. W. (2005). *Masculinities* (2nd ed.). Berkeley: University of California Press.

Connell, R. W., & Messerschmidt, J. W. (2005). Hegemonic masculinity: Rethinking the concept. *Gender & Society, 19*(6), 829–859.

Consalvo, M. (2007). *Cheating: Gaining advantage in video games.* Cambridge, MA: The MIT Press.

———. (2009). There is no magic circle. *Games and Culture, 4*(4), 408–417.

Coscarelli, J. (2012, April 12). Is Reddit being scared straight for encouraging a "suicide"? *New York.* Retrieved April 30, 2014, from http://nymag.com/daily/intelligencer/2012/04/reddit-scared-straight-for-encouraging-suicide.html

Costikyan, G. (2002). I Have No Words & I Must Design: Toward a Critical Vocabulary for Games. In F. Mäyrä (Ed.), *Proceedings of Computer Games and Digital Cultures Conference.* Tampere: Tampere University Press.

creep_creepette. (2013, March 2). Today this chick said some shit like "There are some things that are known which can't be learned."—What the fuck kind of Epistemology is this shit? Retrieved March 10, 2014, from http://www.reddit.com/r/fuckingphilosophy/comments/19jrhc/today_this_chick_said_some_shit_like_there_are/

Culp-Ressler, T. (2013, December 20). "Men's rights" groups are spamming Occidental College with hundreds of false rape reports. *Think Progress.* Retrieved April 5, 2014, from http://thinkprogress.org/health/2013/12/20/3093761/mens-rights-occidental/

cupcake1713. (2014a, June). Frequently Asked Questions—How is a submissions's score determined? Retrieved July 5, 2014, from http://www.reddit.com/wiki/faq - wiki_how_is_a_submission.27s_score_determined.3F

———. (2014b, February 24). remember the human. Retrieved June 27, 2014, from http://www.reddit.com/r/blog/comments/1ytp7q/remember_the_human/

Dachis, A. (2012, January 18). All about PIPA and SOPA, the bills that want to censor your internet. Retrieved March 10, 2014, from http://lifehacker.com/5860205/all-about-sopa-the-bill-thats-going-to-cripple-your-internet

Dao_of_Tao. (2012, September 23). Okay /r/cringe, we need to have a talk. Retrieved April 14, 2014, from http://www.reddit.com/r/cringe/comments/10cg0c/okay_rcringe_we_need_to_have_a_talk

Davison, P. (2012). The language of internet memes. In M. Mandiberg (Ed.), *The social media reader* (pp. 120–134). New York: NYU Press.

definitelymyrealname. (2013, June 23). What's the drama surrounding Game of Trolls? Retrieved July 16, 2014, from http://www.reddit.com/r/OutOfTheLoop/comments/1gwgmk/whats_the_drama_surrounding_game_of_trolls/

Deimorz. (2014a, May 23). It's been two weeks since TwoX became a default [Comment]. Retrieved June 6, 2014, from http://www.reddit.com/r/TwoXChromosomes/comments/26b-8fz/its_been_two_weeks_since_twox_became_a_default/chpmak7

———. (2014b, June 18). reddit changes: individual up/down vote counts no longer visible, "% like it" closer to reality, major improvements to "controversial" sorting. Retrieved July 7, 2014, from http://www.reddit.com/r/announcements/comments/28hjga/reddit_changes_individual_updown_vote_counts_no/

Delwiche, A., & Henderson, J. J. (Eds.). (2013). *The participatory cultures handbook*. New York: Routledge.

devicerandom. (2012, January 15). A summary of reddit cosmology or, on the forum as a work of art. Retrieved May 29, 2014, from http://blog.devicerandom.org/2012/01/15/reddit-cosmology/

dhamster. (2014, February 24). Remember the karma. Retrieved June 27, 2014, from http://www.reddit.com/r/circlejerk/comments/1ytzeh/remember_the_karma/

Dibbell, J. (1999). *My tiny life: Crime and passion in a virtual world*. New York: Henry Holt.

DingoQueen. (2014, January 31). She sat at the shelter for over 3 weeks, showing off her underbite, and trying to get a new mom. Well ... Now she's mine! Retrieved April 5, 2014, from http://www.reddit.com/r/aww/comments/1wp5dr/she_sat_at_the_shelter_for_over_3_weeks_showing/

Dixon-Thayer, D. (2014, January 18). Remebering SOPA/PIPA [Blog post]. *The Mozilla Blog*. Retrieved March 10, 2014, from https://blog.mozilla.org/blog/2014/01/18/remembering-sopapipa/

djloreddit. (2014, February 18). What is the best way to sort? Top? Best? New? Retrieved July 10, 2014, from http://www.reddit.com/r/TheoryOfReddit/comments/1y8rst/what_is_the_best_way_to_sort_top_best_new/

djmushroom. (2014, May 30). Prepared a teddy bear gift yesterday for my gf, and now it looks like this. Retrieved May 30, 2014, from http://www.reddit.com/r/funny/comments/26ug5d/prepared_a_teddy_bear_gift_yesterday_for_my_gf/

domo-loves-yoshi. (2014, January 16). Just bought this today, will be my first time playing. . Retrieved May 28, 2014, from http://www.reddit.com/r/wow/comments/1vdetx/just_bought_this_today_will_be_my_first_time/

Don. (2013a, December). Maury Lie Detector. *Know Your Meme*. Retrieved April 20, 2014, from http://knowyourmeme.com/memes/maury-lie-detector

———. (2013b). White Knight. *Know Your Meme*. Retrieved May 4, 2014, from http://knowyourmeme.com/memes/white-knight

Donald Glover—Crazy stories. (2010). *Comedy Central presents*. Retrieved April 11, 2014, from http://www.cc.com/video-clips/jm3dvg/comedy-central-presents-crazy-stories

Donath, J. S. (1999). Identity and deception in the virtual community. In M. A. Smith & P. Kollock (Eds.), *Communities in cyberspace*. London: Routledge.

dont_stop_smee. (2013, March 16). A friend of mine moved into a former drug house and found this HUGE safe. How do we get it open? Retrieved July 3, 2014, from http://www.reddit.com/r/pics/comments/1aenk5/a_friend_of_mine_moved_into_a_former_drug_house/

Down-n-Dirty. (2013, June 8). What kind of bullshit is this? Retrieved April 6, 2014, from http://www.reddit.com/r/MensRights/comments/1fyeh3/what_kind_of_bullshit_is_this/

Duggan, M., & Smith, A. (2013, July 3). 6% of online adults are reddit users. *PewResearch Internet Project*. Retrieved March 10, 2014, from http://www.pewinternet.org/2013/07/03/6-of-online-adults-are-reddit-users/

Dunn, G. (2013, May 24). The dark side of Reddit's GoneWild. *The Daily Dot*. Retrieved April 5, 2014, from http://www.dailydot.com/society/reddit-gone-wild-dark-side/

During summer, is there actually an increase of young using Reddit, or a noticeable drop in quality? (2013, September 6). Retrieved June 24, 2014, from http://www.reddit.com/r/TheoryOfReddit/comments/1lvqad/during_summer_is_there_actually_an_increase_of/

Dym, C. L., Agogino, A. M., Eris, O., Frey, D. D., & Leifer, L. J. (2005). Engineering design thinking, teaching, and learning. *Journal of Engineering Education*, 94(1), 103–120. doi: 10.1002/j.2168-9830.2005.tb00832.x

EffinD. (2011, August 14). Shower to go. Retrieved August 1, 2013, from http://www.reddit.com/r/pics/comments/jinex/shower_to_go/

Ellison, N. B., & boyd, d. m. (2013). Sociality through social network sites. In W. H. Dutton (Ed.), *The Oxford handbook of internet studies* (pp. 151–172). Oxford: Oxford University Press.

EMCoupling. (2014, April 17). Dear Confession Bear guy: [Comment]. Retrieved April 20, 2014, from http://www.reddit.com/r/AdviceAnimals/comments/23axet/dear_confession_bear_guy/cgvcy1u

Emerson, R. M., Fretz, R. I., & Shaw, L. L. (2011). *Writing ethnographic fieldnotes* (2nd ed.). Chicago: The University of Chicago Press.

Engeström, Y. (1999). Activity theory and transformation. In Y. Engeström, R. Miettinen, & R.-L. Punamäki-Gitai (Eds.), *Perspectives on activity theory* (pp. 19–38). Cambridge: Cambridge University Press.

Engeström, Y., Miettinen, R., & Punamäki-Gitai, R.-L. (1999). *Perspectives on activity theory*. Cambridge: Cambridge University Press.

Ensmenger, N. (2010). *The computer boys take over: Computers, programmers, and the politics of technical expertise*. Cambridge, MA: The MIT Press.

faq—MensRights. (2014, February). Retrieved April 5, 2014, from http://www.reddit.com/r/MensRights/wiki/faq

Flanagan, M. (2009). *Critical play: Radical game design*. Cambridge, MA: The MIT Press.

flobbley. (2013, March 7). This rock cracked open on my construction site today. Retrieved August 11, 2013, from http://www.reddit.com/r/pics/comments/19voto/this_rock_cracked_open_on_my_construction_site/

Foot, K., & Groleau, C. (2011). Contradictions, transitions, and materiality in organizing processes: An activity theory perspective. *First Monday*. Retrieved July 14, 2014, from http://firstmonday.org/ojs/index.php/fm/article/view/3479/2983

ForWhatReason. (2011, November 4). ELI5 "The Great Digg Migration." Retrieved June 2, 2014, from http://www.reddit.com/r/explainlikeimfive/comments/m0w30/eli5_the_great_digg_migration/

Frank_Reynolds_19. (2014, May 12). I frequently see this comment on Reddit and it makes me rage every time. Retrieved June 30, 2014, from http://www.reddit.com/r/circlebroke/comments/25chwx/very_low_effort_i_frequently_see_this_comment_on/

Fung, K. (2013, April 21). Media criticize *New York Post*, CNN for Boston Marathon bombings coverage. Retrieved March 10, 2014, from http://www.huffingtonpost.com/2013/04/21/media-boston-new-york-post-cnn_n_3127883.html

G.F. (2012, April 9). Eternal September lives on. *The Economist (Babbage Blog)*. Retrieved June 2, 2014, from http://www.economist.com/blogs/babbage/2012/04/internet-mores

Galloway, A. R. (2006). *Gaming: Essays on algorithmic culture*. Minneapolis: University of Minnesota Press.

genius-bar. (2014, June 8). Pizza insecurities. Retrieved June 29, 2014, from http://www.reddit.com/r/SubredditDrama/comments/27mis9/pizza_insecurities/

Gillespie, T. (2010). The politics of "platforms." *New Media & Society*, *12*(3), 347–364.

———. (2014). The relevance of algorithms. In T. Gillespie, P. J. Boczkowski, & K. A. Foot (Eds.), *Media technologies: Essays on communication, materality, and society* (pp. 167–194). Cambridge, MA: The MIT Press.

Gillespie, T., Boczkowski, P. J., & Foot, K. A. (2014). Introduction. In T. Gillespie, P. J. Boczkowski, & K. A. Foot (Eds.), *Media technologies: Essays on communication, materality, and society* (pp. 1–17). Cambridge, MA: The MIT Press.

Gillmor, D. (2006). *We the media: Grassroots journalism by the people, for the people*. Sebastopol, CA: O'Reilly Media.

girrrrrrrrrl. (2012, October 4). Why does the reddit search engine suck so bad? Retrieved July 16, 2014, from http://www.reddit.com/r/explainlikeimfive/comments/10z9yr/why_does_the_reddit_search_engine_suck_so_bad

Goffman, E. (1959). *The presentation of self in everyday life*. Garden City, NY: Doubleday.

goldfish188. (2012, October 8). May I ask exactly what the point of this subreddit is? Retrieved March 10, 2014, from http://www.reddit.com/r/Music/comments/114rfm/may_i_ask_exactly_what_the_point_of_this/

goldguy81. (2014, June 25). How much does your reddit account reflect your personality? Retrieved June 27, 2014, from http://www.reddit.com/r/TheoryOfReddit/comments/292i2b/how_much_does_your_reddit_account_reflect_your/

Granovetter, M. S. (1983). The strength of weak ties: A network theory revisited. *Sociological Theory*, *1*, 203–233.

Gray, K. L. (2012). Intersecting oppressions and online communities: Examining the experiences of women of color in Xbox Live. *Information, Communication & Society*, *15*(3), 411–428. doi: 10.1080/1369118X.2011.642401

Gray, M. (2012). *CAUTION!! Boundary work ahead for internet studies ... or, Why the twilight of the "toaster studies" approach to internet research is a very, very good thing*. Paper presented at the Association of Internet Researchers Annual Conference—IR13, Salford, UK. http://socialmediacollective.org/2012/12/17/caution-boundary-work-ahead-for-internet-studies/

Greenberg, A. (2014, September 10). Hacked celeb pics made reddit enough cash to run its servers for a month. *Wired*. Retrieved September 10, 2014, from http://www.wired.com/2014/09/celeb-pics-reddit-gold/

Halavais, A. (2009). Do Dugg Diggers Digg Diligently? Feedback as motivation in collaborative moderation systems. *Information, Communication & Society, 12*(3), 444–459.

hansjens47. (2014, April). /r/Politics wiki—rulesandregs. Retrieved June 23, 2014, from http://www.reddit.com/r/politics/wiki/rulesandregs

happybadger. (2011, November 9). /Out of Character/ A proposition based on the recent discussion in r/fifthworldproblems [Comment]. Retrieved May 28, 2014, from http://www.reddit.com/r/sixthworldproblems/comments/m70em/out_of_character_a_proposition_based_on_the/c2z4ujm

Hardin, G. (1968). The tragedy of the commons. *Science, 162*(3859), 1243–1248.

Haughney, C. (2013, June 6). *New York Post* faces suit over Boston Bomb article. Retrieved March 10, 2014, from http://www.nytimes.com/2013/06/07/business/media/new-york-post-sued-over-boston-bombing-article.html

Haywood, B. (2012, May 24). Hipster racism: "Irony" and the lingering cultural capital of racism. *Left of Center*. Retrieved June 6, 2014, from http://brittspolitical.wordpress.com/2012/05/24/hipster-racism-irony-the-lingering-cultural-capital-of-race-and-racism/

HeLivesMost. (2014, April 17). Dear confession bear guy:. Retrieved April 20, 2014, from http://www.reddit.com/r/AdviceAnimals/comments/23axet/dear_confession_bear_guy/

Henricks, T. S. (2006). *Play reconsidered: Sociological perspectives on human expression*. Urbana: University of Illinois Press.

Hern, A. (2013, April 19). When crowdsourcing goes wrong: Reddit, Boston and missing student Sunil Tripathi. *New Statesman*. Retrieved March 10, 2014, from http://www.newstatesman.com/world-affairs/2013/04/reddit-boston-and-missing-student

———. (2014, May 8). Reddit shakes up the defaults again: which are the best of the new intake? *The Guardian*. Retrieved July 15, 2014, from http://www.theguardian.com/technology/2014/may/08/reddit-defaults-which-are-the-best

highshelfofsteam. (2014, February 26). One hundred gift exchanges: A celebration extravaganzapalooza. Retrieved May 10, 2014, from http://redditgifts.com/blog/view/one-hundred-gift-exchanges-celebration-extravaganzapalooza/

Hilarious_Exception. (2013, December 29). One of the Best Gifts I Have Received Ever. Retrieved May 10, 2014, from http://redditgifts.com/gallery/gift/one-best-gifts-have-received-ever/

Hindman, M. (2008). *The myth of digital democracy*. Princeton, NJ: Princeton University Press.

Hine, C. (2008). Virtual ethnography: Modes, varieties, affordances. In N. G. Fielding, R. M. Lee, & G. Blank (Eds.), *The Sage handbook of online research methods* (pp. 257–271). Thousand Oaks, CA: Sage.

hitbart000. (2014, April 20). For the third time in two years, feminists at a major Canadian university interrupt a male issues lecture on campus. The feminists yelled and talked over the lecturer, blew air horns in the lecture room, and later pulled a fire alarm to sabotage the lecture.—[1:32]. Retrieved April 20, 2014, from http://www.reddit.com/r/videos/comments/23ij7t/for_the_third_time_in_two_years_feminists_at_a/

hmasing. (2010, October 11). REDDIT—Kathleen, the little girl who was being harassed by her neighbors, has a special message for you. Retrieved May 14, 2014, from http://www.reddit.com/r/reddit.com/comments/dpwiu/reddit_kathleen_the_little_girl_who_was_being/

Hogan, B. (2013). Pseudonyms and the rise of the real-name web. In J. Hartley, J. Burgess, & A. Bruns (Eds.), *A companion to new media dynamics* (pp. 290–308). New York: Wiley-Blackwell.

Holcomb, J., Gottfried, J., & Mitchell, A. (2013). *News use across social media platforms*. Pew Research Center. Retrieved from http://www.journalism.org/files/2013/11/News-Use-Across-Social-Media-Platforms1.pdf

hooks, b. (1992). *Black looks: Race and representation*. Cambridge, MA: South End Press.

hoontur. (2013, June 19). What do you guys do with almond meal? Retrieved August 1, 2013, from http://www.reddit.com/r/apple/comments/1hbbf2/what_do_you_guys_do_with_almond_meal/

Horst, H. A., & Miller, D. (Eds.). (2012). *Digital anthropology*. London: Bloomsbury Academic.

Hp9rhr. (2014, March 22). Which is your favourite reddit bot and how can it be summoned? Retrieved May 21, 2014, from http://www.reddit.com/r/AskReddit/comments/212zxo/which_is_your_favourite_reddit_bot_and_how_can_it/

hth6565. (2014, May 28). To the people that said the design wouldn't go through—here is a picture as promised (OP delivers!) [Comment]. Retrieved May 28, 2014, from http://www.reddit.com/r/funny/comments/26p1dd/to_the_people_that_said_the_design_wouldnt_go/chta4af?context=1

Hughey, M. W., & Daniels, J. (2013). Racist comments at online news sites: A methodological dilemma for discourse analysis. *Media, Culture & Society, 35*(3), 332–347. doi: 10.1177/0163443712472089

Huizinga, J. (1971). *Homo ludens: A study of the play element in culture*. Boston: Beacon Press.

Humphreys, S., & Vered, K. O. (2014). Reflecting on gender and digital networked media. *Television & New Media, 15*(3), 3–13. doi: 10.1177/1527476413502682

I_am_one_wth_it_all. (2014, April 17). To the college student who felt that he "wasted" four years of his life. Retrieved April 18, 2014, from http://www.reddit.com/r/AdviceAnimals/comments/239hqa/to_the_college_student_who_felt_that_he_wasted/

iamwoodyharrelson. (2012, February 3). I'm Woddy Harrelson, AMA. Retrieved April 20, 2014, from http://www.reddit.com/r/IAmA/comments/p9a1v/im_woody_harrelson_ama/c3nmf7t

Imafidon, A.-M. (2013, May 9). Are all-male panels at tech conferences a thing of the past? *The Guardian*. Retrieved April 1, 2014, from http://www.theguardian.com/women-in-leadership/2013/may/09/all-male-panels-tech-conferences

Ireallylikepbr. (2014, March 29). This just came out of a coworkers mouth. Retrieved April 5, 2014, from http://www.reddit.com/r/AdviceAnimals/comments/21n117/this_just_came_out_of_a_coworkers_mouth/

REFERENCES

JackRig95. (2013, July 14). Girl, you need new sneakers! Retrieved April 11, 2014, from http://www.reddit.com/r/CandidFashionPolice/comments/1ia4ns/girl_you_need_new_sneakers

James. (2014, March 7). Socially Awkward Penguin. Retrieved March 10, 2014, from http://knowyourmeme.com/memes/socially-awkward-penguin

jasoneppink. (2014, February 25). Hey Reddit, want to help curate a museum exhibition about reaction gifs? (details in comments). Retrieved May 22, 2014, from http://www.reddit.com/r/gifs/comments/1yw7aa/hey_reddit_want_to_help_curate_a_museum/

jedberg. (2010, November 23). jedberg comments on pardon me, but 5000 downvotes? WTF is "worldnews" for??? Retrieved August 1, 2013, from http://www.reddit.com/r/WTF/comments/eaqnf/pardon_me_but_5000_downvotes_wtf_is_worldnews_for/c16omup

Jenkins, H. (2006). *Convergence culture: Where old and new media collide.* New York: NYU Press.

Jenkins, H., Ford, S., & Green, J. (2013). *Spreadable media: Creating value and meaning in a networked world.* New York: NYU Press.

Jenkins, H., Purushotma, R., Weigel, M., Clinton, K., & Robison, A. J. (2009). *Confronting the challenges of participatory culture: Media education for the 21st century.* Cambridge, MA: The MIT Press.

jkerwin. (2007, February 19). The Eternal September—a good analogy for what's happened to Digg and is starting to happen to reddit. Retrieved June 2, 2014, from http://www.reddit.com/r/programming/comments/15j08/the_eternal_september_a_good_analogy_for_whats/

jmk4422. (2011, October 11). October 12, 2011. /r/shitredditsays. Airing our dirty laundry in public. Retrieved May 5, 2014, from http://www.reddit.com/r/subredditoftheday/comments/l969q/october_12_2011_rshitredditsays_airing_our_dirty/

JohnArr. (2013, July 29). The saddest cookbook. Retrieved August 11, 2013, from http://www.reddit.com/r/funny/comments/1j9wq9/the_saddest_cookbook/

Johnmtl. (2010, December 3). Thank you! Retrieved May 10, 2014, from http://redditgifts.com/gallery/gift/thanks-you/

Jones, S. G. (Ed.). (1997). *Virtual culture: Identity and communication in cybersociety.* London: Sage.

———. (Ed.). (1998). *Cybersociety 2.0: Revisiting computer-mediated community and technology.* London: Sage.

joserskid. (2014, January 1). My girlfriend pulled this one on me today. Retrieved April 6, 2014, from http://www.reddit.com/r/AdviceAnimals/comments/1u6mi8/my_girlfriend_pulled_this_one_on_me_today/

jrivers242. (2014, March 9). Redditor's Wife. Retrieved April 11, 2014, from http://www.reddit.com/r/AdviceAnimals/comments/1zyqr8/redditors_wife/

judgementbarandgrill. (2012, November 14). A plea to my fellow cringers. Retrieved April 15, 2014, from http://www.reddit.com/r/cringe/comments/136yim/a_plea_to_my_fellow_cringers/

Jurgenson, N. (2011, February 24). Digital dualism versus augmented reality. *Cyborgology.* Retrieved July 10, 2014, from http://thesocietypages.org/cyborgology/2011/02/24/digital-dualism-versus-augmented-reality/

Juul, J. (2005). *Half-real: Video games between real rules and fictional worlds.* Cambridge, MA: The MIT Press.

Kang, J. C. (2013, July 25). Should reddit be blamed for the spreading of a smear? *The New York Times Magazine*. Retrieved July 15, 2014, from http://www.nytimes.com/2013/07/28/magazine/should-reddit-be-blamed-for-the-spreading-of-a-smear.html

Kaptelinin, V., & Nardi, B. A. (2006). *Acting with technology: Activity theory and interaction design*. Cambridge, MA: The MIT Press.

karmarank. (2014, June 12). Karma inequality: 1% of redditors have 20% of the comment karma. Retrieved July 5, 2014, from http://www.reddit.com/r/dataisbeautiful/comments/27zyh6/karma_inequality_1_of_redditors_have_20_of_the/

Kelemen, M., & Smith, W. (2001). Community and its "virtual" promises: A critique of cyber-libertarian rhetoric. *Information, Communication & Society*, 4(3), 370–387. doi: 10.1080/713768547

Kendall, L. (2002). *Hanging out in the virtual pub: Masculinities and relationships online*. Berkeley: University of California Press.

Kerckhove, C. V. (2007, January 15). The 10 biggest race and pop culture trends of 2006: Part 1 of 3. *Racialicious*. Retrieved June 2, 2014, from http://www.racialicious.com/2007/01/15/the-10-biggest-race-and-pop-culture-trends-of-2006-part-1-of-3/

Kid attempts to twerk in his private school uniform to "Birthday Cake." (2012, November 27). Retrieved May 10, 2014, from http://www.reddit.com/r/cringe/comments/13vi7r/kid_attempts_to_twerk_in_his_private_school/

KidsAre9532. (2014, May 18). What is something that screams douchebag from a person's appearance? [Comment]. Retrieved May 21, 2014, from http://www.reddit.com/r/AskReddit/comments/25v56j/what_is_something_that_screams_douchebag_from_a/chl9v9u

Kim, B. (2013, July 1). Know Your Meme: Confession Bear. Retrieved July 1, 2013, from http://knowyourmeme.com/memes/confession-bear

Kimmel, M. (2013). *Angry white men: American masculinity at the end of an era*. New York: Nation Books.

Kirkpatrick, D. (2010). *The Facebook effect: The inside story of the company that is connecting the world*. New York: Simon & Schuster.

KittyFooties. (2014, April 17). Dear Confession bear guy: [Comment]. Retrieved April 20, 2014, from http://www.reddit.com/r/AdviceAnimals/comments/23axet/dear_confession_bear_guy/cgvcdby

kn0thing. (2010, January 14). Helping Haiti (because we ought to do more than just change our logo). Retrieved May 14, 2014, from http://www.reddit.com/r/blog/comments/apnsu/helping_haiti_because_we_ought_to_do_more_than/

Knuttila, L. (2011, October). User unknown: 4chan, anonymity and contingency. *First Monday*. Retrieved January 1, 2013, from http://firstmonday.org/htbin/cgiwrap/bin/ojs/index.php/fm/article/view/3665/3055

Kollock, P., & Smith, M. (Eds.). (1999). *Communities in cyberspace*. London: Routledge.

Kosnik, A. D. (2013). Fandom as free labor. In T. Scholz (Ed.), *Digital labor: The internet as playground and factory* (pp. 98–111). New York: Routledge.

lacylola. (2011, December 12). Never in my wildest dreams. Retrieved May 10, 2014, from http://redditgifts.com/gallery/gift/never-my-wildest-dreams/

ladyofchaos. (2011, December 16). My Santa wrote me a song. Sorry, did I say song? I meant FOUR-PART CHORAL ARRANGEMENT OF MY FAVOURITE POEM. Retrieved May 10, 2014, from http://redditgifts.com/gallery/gift/my-favourite-poem-turned-four-part-piece/

Laina. (Producer). (2012, June 6). JB Fanvideo. [Video] Retrieved from https://http://www.youtube.com/watch?v=Yh0AhrY9GjA

LainaOAG. (2012, September 28). I'm Laina AKA "Overly Attached Girlfriend" AMA. Retrieved April 11, 2014, from http://www.reddit.com/r/AdviceAnimals/comments/10mnfy/im_laina_aka_overly_attached_girlfriend_ama/

Lanier, J. (2010). *You are not a gadget*. New York: Alfred A. Knopf.

laptopdude90. (2014, May 4). Extra_Cheer_Bot: Cheers you up if you're feeling down! Retrieved May 20, 2014, from http://www.reddit.com/r/botwatch/comments/24ow7e/extra_cheer_bot_cheers_you_up_if_youre_feeling/

Lardinois, F. (2010, August 30). Digg user rebellion continues: Reddit now rules the front page. *ReadWrite*. Retrieved June 2, 2014, from http://readwrite.com/2010/08/30/digg_user_rebellion_reddit_on_front_page - awesm=~oG8yijUja2TP8A

LaTeX_fetish. (2012, November 26). [Effort] What makes Gina a good girl? Retrieved April 6, 2014, from http://www.reddit.com/r/ShitRedditSays/comments/13ufh3/effort_what_makes_gina_a_good_girl/

Latour, B. (1988). *Science in action: How to follow scientists and engineers through society*. Cambridge, MA: Harvard University Press.

———. (1992). Where are the missing masses? The sociology of a few mundane artifacts. In W. E. Bijker & J. Law (Eds.), *Shaping technology/building society* (pp. 225–257). Cambridge, MA: The MIT Press.

———. (2005). *Reassembling the social: An introduction to actor-network-theory*. Oxford: Oxford University Press.

LavaLampJuice. (2013, December 12). My current girlfriend. Retrieved April 5, 2014, from http://www.reddit.com/r/AdviceAnimals/comments/1sq1bg/my_current_girlfriend/

Lave, J., & Wenger, E. (1991). *Situated learning: Legitimate peripheral participation*. New York: Cambridge University Press.

Lerman, K., & Ghos, R. (2010). *Information contagion: An empirical study of the spread of news on Digg and Twitter social networks*. Paper presented at the Fourth International AAAI Conference on Weblogs and Social Media, Washington, DC.

Levy, S. (2010). *Hackers: Heroes of the computer revolution—25th Anniversary edition*. Sebastopol, CA: O'Reilly Media.

Lewis, H. (2012, June 12). Dear the internet, This is why you can't have anything nice. *New Statesman*. Retrieved April 1, 2014, from http://www.newstatesman.com/blogs/internet/2012/06/dear-internet-why-you-cant-have-anything-nice

LGBTerrific. (2013, August 14). August 14, 2013—/r/switcharoo—Going deeper into reddit than you ever wanted to Retrieved May 28, 2014, from http://www.reddit.com/r/subredditoftheday/comments/1kc3gg/august_14_2013_rswitcharoo_going_deeper_into/

Liebelson, D., & Raja, T. (2013, March 22). Donglegate: How one brogrammer's sexist joke led to death threats and firings. *Mother Jones*. Retrieved April 1, 2014, from http://www.motherjones.com/politics/2013/03/pycon-2013-sexism-dongle-richards

Lindell. (2010). There Are No Girls On The Internet. *Know Your Meme*. Retrieved April 20, 2014, from http://knowyourmeme.com/memes/there-are-no-girls-on-the-internet

LogicPlacebo. (2014, April 17). As a college student about to graduate ... Retrieved April 18, 2014, from http://www.reddit.com/r/AdviceAnimals/comments/238lli/as_a_college_student_about_to_graduate/

lukeis2cool. (2013, December 19). My ex regrets going to the Child Support Agency (CSA) now, I'm due a refund (UK). Retrieved April 5, 2014, from http://www.reddit.com/r/AdviceAnimals/comments/1t8mtj/my_ex_regrets_going_to_the_child_support_agency/

Lynn, A. (2013, January 7). A primer on the "Friend Zone." *Nerdy Feminist*. Retrieved April 5, 2014, from http://www.nerdyfeminist.com/2013/01/a-primer-on-friend-zone.html

———. (2014, May 6). When did we lose our understanding of "satire"? Retrieved September 25, 2014, from http://www.nerdyfeminist.com/2014/05/when-did-we-lose-our-understanding-of.html

MacFarquar, L. (2013, March 11). Requiem for a dream. *New Yorker*. Retrieved July 5, 2014, from http://www.newyorker.com/reporting/2013/03/11/130311fa_fact_macfarquhar

MacTheMan. (2013, March 25). Ayn Rand what the fuck you say?!?!?!?! Retrieved 2014, March 10, from http://www.reddit.com/r/fuckingphilosophy/comments/1b0t06/ayn_rand_what_the_fuck_you_say/

Madrigal, A. (2014, 7 Jan). AMA: How a weird internet thing became a mainstream delight. *The Atlantic*. Retrieved January 13, 2014, from http://www.theatlantic.com/technology/archive/2014/01/ama-how-a-weird-internet-thing-became-a-mainstream-delight/282860/

Marin, A. (2012, November 28). Overly attached girlfriend Laina Walker: From meme to TV presenter. *Policy Mic*. Retrieved April 11, 2014, from http://www.policymic.com/articles/19763/overly-attached-girlfriend-laina-walker-from-meme-to-tv-presenter

Markham, A. N. (2009). How can qualitative researchers produce work that is meaningful across time, space, and culture? In A. N. Markham & N. K. Baym (Eds.), *Internet inquiry: Conversations about method* (pp. 131–155). Thousand Oaks, CA: Sage.

Markham, A. N., & Baym, N. K. (Eds.). (2009). *Internet inquiry: Conversations about method*. Thousand Oaks, CA: Sage.

Marlow, C., Naaman, M., Boyd, D., & Davis, M. (2006). *HT06, tagging paper, taxonomy, Flickr, academic article, to read*. Paper presented at the 17th conference on Hypertext and Hypermedia, Odense, Denmark.

Marwick, A. E. (2013). *Status update: Celebrity, publicity, and branding in the social media age*. New Haven, CT: Yale University Press.

Marwick, A. E., & boyd, d. m. (2011a). *The drama! Teen conflict, gossip, and bullying in networked publics*. Paper presented at the Oxford Internet Institute's "A Decade in Internet Time: Symposium on the Dynamics of the Internet and Society." Retrieved from http://ssrn.com/abstract=1926349

———. (2011b). I tweet honestly, I tweet passionately: Twitter users, context collapse, and the imagined audience. *New Media & Society, 13*(1), 114–133. doi: 10.1177/1461444810365313

Mathis, G. (2013, February 5). Funds raised for Atlanta security guard who "tased" woman. *News to Me with George Mathis*. Retrieved May 15, 2014, from http://blogs.ajc.com/news-to-me/2013/02/05/funds-raised-for-atlanta-security-guard-who-tased-woman/

maxgprime. (2014, May 5). We're fighting for marriage equality in Utah and around the world. Will you help us? Retrieved June 20, 2014, from http://www.reddit.com/r/blog/comments/24seva/were_fighting_for_marriage_equality_in_utah_and/

Maxion. (2013, September 7). Reddit is no longer a link aggregator, it is an image board. Retrieved May 23, 2014, from http://www.reddit.com/r/TheoryOfReddit/comments/zhy66/reddit_is_no_longer_a_link_aggregator_it_is_an/

McArthur, J. A. (2009). Digital subculture: A geek meaning of style. *Journal of Communication Inquiry, 33*(1), 58–70.

McCombs, M. E., & Shaw, D. L. (1972). The agenda-setting function of mass media. *Public Opinion Quarterly, 36*(2), 176–187. doi: 10.1086/267990

Mcodray. (2013, February 11). Redditor's wife fooled three days before valentine day. Retrieved April 11, 2014, from http://www.reddit.com/r/AdviceAnimals/comments/18cwb2/redditors_wife_fooled_three_days_before_valentine/

Meyer, R. (2014, June 28). Everything we know about Facebook's secret mood manipulation experiment. *The Atlantic*. Retrieved July 22, 2014, from http://www.theatlantic.com/technology/archive/2014/06/everything-we-know-about-facebooks-secret-mood-manipulation-experiment/373648/

mikey_mike24. (2013, July 12). Reddit, what overused phrase used on this site do you hate the most? Retrieved July 1, 2014, from http://www.reddit.com/r/AskReddit/comments/1i5cd2/reddit_what_overused_phrase_used_on_this_site_do/

Milner, R. M. (2009). Working for the text: Fan labor and the New Organization. *International Journal of Cultural Studies, 12*(5), 491–508.

———. (2012). *The world made meme: Discourse and identity in participatory media* (Doctoral dissertation). University of Kansas, Lawrence, KS. Retrieved from http://www.academia.edu/2065427/The_World_Made_Meme_Discourse_and_Identity_in_Participatory_Media

Misa, T. J. (2010). Gender codes: Lessons from history. In T. J. Misa (Ed.), *Gender codes: Why women are leaving computing* (pp. 251–264). Hoboken, NJ: John Wiley & Sons.

MittRomneysCampaign. (2013, February 26). ShitRedditSays: FAQ. Retrieved May 8, 2014, from http://www.reddit.com/r/SRSsucks/comments/19a9z9/shitredditsays_faq/

moab-girl. (2014, February 27). She's 23 and goes out in public like this. Retrieved July 15, 2014, from http://www.reddit.com/r/cringepics/comments/1z3hqc/shes_23_and_goes_out_in_public_like_this/

Montfort, N. (2014, July 13). The facepalm at the end of the mind. *Post Position*. Retrieved July 19, 2014, from http://nickm.com/post/2014/07/the-facepalm-at-the-end-of-the-mind/

Moore, C. (2011). The magic circle and the mobility of play. *Convergence: The International Journal of Research into New Media Technologies, 17*(4), 373–387. doi: 10.1177/1354856511414350

Morozov, E. (2013). *To save everything, click here: The folly of technological solutionism*. New York: PublicAffairs.

Morris, K. (2012a, March 23). How r/ShitRedditSays became Reddit's bomb-throwing counter-culture. Retrieved January 10, 2014, from http://www.dailydot.com/society/reddit-pedo-geddon-shitredditsays/

———. (2012b, July 26). Reddit bans its biggest troll hub. *The Daily Dot*. Retrieved July 17, 2014, from http://www.dailydot.com/news/reddit-digest-gameoftrolls-banned/

Mowbray, M. (2014). Automated Twitter accounts. In K. Weller, A. Bruns, J. Burgess, M. Mahrt, & C. Puschmann (Eds.), *Twitter and society* (pp. 183–194). New York: Peter Lang.

Muchnik, L., Aral, S., & Taylor, S. J. (2013). Social influence bias: A randomized experiment. *Science, 341*, 647–651. doi: 10.1126/science.1240466

Mulvey, L. (1989). *Visual and other pleasures*. Bloomington: Indiana University Press.

Munroe, R. (2009, October 15). Reddit's new comment sorting system [Blog post]. *The Reddit Blog*. Retrieved July 10, 2014, from http://www.redditblog.com/2009/10/reddits-new-comment-sorting-system.html

Murray, S. (2004). "Celebrating the story the way it is": Cultural studies, corporate media, and the contested utility of fandom. *Continuum: Journal of Media & Cultural Studies, 18*(1), 7–25.

MuscleT. (2012, September 29). Redditors wife has bad news. Retrieved April 11, 2014, from http://www.reddit.com/r/AdviceAnimals/comments/10nvaq/redditors_wife_has_bad_news/

Museum of the Moving Image—Exhibitions—The reaction GIF: Moving image as gesture. (2014). Retrieved May 22, 2014, from http://www.movingimage.us/exhibitions/2014/03/12/detail/the-reaction-gif-moving-image-as-gesture/

Mylaptopisburningme. (2013, January 28). Woman gets up in security guards face. Gets a well deserved taser. Retrieved May 14, 2014, from http://www.reddit.com/r/JusticePorn/comments/17firo/woman_gets_up_in_security_guards_face_gets_a_well/

Nahon, K., & Hemsley, J. (2013). *Going viral*. Cambridge: Polity.

Nakamura, L. (2002). *Cybertypes: Race, ethnicity, and identity on the internet*. New York: Routledge.

———. (2007). *Digitizing race: Visual cultures of the internet*. Minneapolis: University of Minnesota Press.

———. (2013, October 23). *Race, gender and information communication technologies*. Paper presented at the Association of Internet Researchers Annual Conference (IR 14), Denver, CO.

Nakamura, L., & Chow-White, P. A. (Eds.). (2012). *Race after the internet*. New York: Routledge.

Nardi, B. A. (1996). *Context and consciousness: Activity theory and human-computer interaction*. Cambridge, MA: The MIT Press.

———. (2010). *My life as a night elf priest: An anthropological account of World of Warcraft*. Ann Arbor: University of Michigan Press.

NetLingo list of chat acronyms & text shorthand. (2014). Retrieved June 6, 2014, from http://www.netlingo.com/acronyms.php

neuroticfish. (2014, March 25). Cringepics user finds his brother on the front page, possibly uploaded by ex. Retrieved March 25, 2014, from http://www.reddit.com/r/SubredditDrama/comments/21ax81/cringepics_user_finds_his_brother_on_the_front/

Nice guy syndrome. (2014, March 15). *Geek Feminism Wiki.* Retrieved April 5, 2014, from http://geekfeminism.wikia.com/wiki/Nice_guy_syndrome

Nicholson, J. (2011, December 16). Escape the Friend Zone: From friend to girlfriend or boyfriend. *Psychology Today.* Retrieved April 5, 2014, from http://www.psychologytoday.com/blog/the-attraction-doctor/201112/escape-the-friend-zone-friend-girlfriend-or-boyfriend

Nicole, K. (2007, December 11). Mr. Splashy Pants and the tale of a hijacked PR campaign. *Mashable.* Retrieved March 10, 2014, from http://mashable.com/2007/12/11/mr-splashy-pants-greenpeace/

Nine things to know about RedditGifts. (2014). Retrieved May 10, 2014, from http://redditgifts.com/about/

Nissenbaum, H. (2010). *Privacy in context: Technology, policy, and the integrity of social life.* Stanford, CA: Stanford University Press.

Norris, P. (2001). *Digital divide: Civic engagement, information poverty, and the internet worldwide.* Cambridge: Cambridge University Press.

Norton, M. I., & Sommers, S. R. (2011). Whites see racism as a zero-sum game that they are now losing. *Perspectives on Psychological Science, 6*(3), 215–218.

NovaXP. (2013). Doge. *Know Your Meme.* Retrieved May 18, 2014, from http://knowyourmeme.com/memes/doge

Nussbaum, M. C. (2010). Objectification and internet misogyny. In S. Levmore & M. C. Nussbaum (Eds.), *The offensive internet: Speech, privacy, and reputation* (pp. 68–87). Cambridge, MA: Harvard University Press.

Nutshapio. (2005, December 12). Reddit now supports comments. Retrieved June 3, 2014, from http://www.reddit.com/r/reddit.com/comments/17913/reddit_now_supports_comments/

O'Reilly, T. (2005, September 30). What is Web 2.0?: Design patterns and business models for the next generation of software. Retrieved June 15, 2009, from http://www.oreillynet.com/pub/a/oreilly/tim/news/2005/09/30/what-is-web-20.html

Ohanian, A. (2013). *Without their permission: How the 21st century will be made, not managed.* New York: Business Plus.

On the theme of higher education haters. (2014, April 17). Retrieved April 18, 2014, from http://www.reddit.com/r/AdviceAnimals/comments/239qqk/on_the_theme_of_higher_education_haters/

Ooer. (2014, June 11). Why does Reddit as a whole single out popular accounts and create a celebrity status around them? Retrieved June 27, 2014, from http://www.reddit.com/r/TheoryOfReddit/comments/27whqp/why_does_reddit_as_a_whole_single_out_popular/

opspe. (2013, April). Rage Comics. *Know Your Meme.* Retrieved August 1, 2013, from http://knowyourmeme.com/memes/rage-comics

OverjoyedMuffin. (2014, June 3). BLACK PEOPLE LIEK WATERMELONS GUYS EDGY AS FUCK. Retrieved July 20, 2014, from http://www.reddit.com/r/SummerReddit/comments/2784ej/black_people_liek_watermelons_guys_edgy_as_fuck/

Pariser, E. (2011). *The filter bubble: What the internet is hiding from you.* New York: Penguin Press.

Pascoe, C. J. (2011). *Dude, you're a fag: Masculinity and sexuality in high school.* Berkeley: University of California Press.

Pearce, C. (2011). *Communities of play: Emergent cultures in multiplayer games and virtual worlds.* Cambridge, MA: The MIT Press.

Pekhota. (2014, June 23). Is Reddit becoming a platform for hateful ideas? Retrieved June 27, 2014, from http://www.reddit.com/r/TheoryOfReddit/comments/28unad/is_reddit_becoming_a_platform_for_hateful_ideas/

pertnear. (2014, February 1). My dad is a plow driver for MNDoT. We had a pretty good snowfall a couple days ago and I got to ride along with him. Here's an album of my experience. Retrieved April 5, 2014, from http://www.reddit.com/r/pics/comments/1wqpyu/my_dad_is_a_plow_driver_for_mndot_we_had_a_pretty/

Phillips, W. (2015). *This is why we can't have nice things: Mapping the relationship between online trolling and mainstream culture.* Cambridge, MA: The MIT Press.

pk_atheist. (2012, November 8). Almost a hundred subscribers! Welcome newcomers. Retrieved April 5, 2014, from http://www.reddit.com/r/TheRedPill/comments/12v1hf/almost_a_hundred_subscribers_welcome_newcomers/

Plague_Bot. (2014, January 28). Bot List: I built a bot to find other bots. So far I have 169 to share with you. Retrieved May 21, 2014, from http://www.reddit.com/r/botwatch/comments/1wg6f6/bot_list_i_built_a_bot_to_find_other_bots_so_far/

PlantSomeTrees. (2014, May 2). I've planted 1.25 Million trees across Canada in the past 8 years, IAMA Treeplanter. Retrieved June 25, 2014, from http://www.reddit.com/r/IAmA/comments/24jgsy/ive_planted_125_million_trees_across_canada_in/

plasmatron7. (2011, October 1). [META] The Best of Pedogeddon. Retrieved May 5, 2014, from http://www.reddit.com/r/ShitRedditSays/comments/kxkip/meta_the_best_of_pedogeddon/

plexico_mcbean. (2013). Dawg, Diogenes was punk as fuck. Retrieved March 10, 2014, from http://www.reddit.com/r/fuckingphilosophy/comments/1ior5i/dawg_diogenes_was_punk_as_fuck/

Popper, B. (2013, April 18). Online witch hunt for Boston Bomber leads to *NY Post* cover photo of innocent "suspects." *The Verge.* Retrieved March 10, 2014, from http://www.theverge.com/2013/4/18/4238768/online-witch-hunt-for-boston-bomber-leads-to-ny-post-cover-photo-mess

Postigo, H. (2010). Modding to the big leagues: Exploring the space between modders and the game industry. *First Monday, 15*(5).

project8an. (2013, August 27). Ran into this one a while back ... Retrieved April 6, 2014, from http://www.reddit.com/r/AdviceAnimals/comments/1l7lll/ran_into_this_one_a_while_back/

/r/antiSRS FAQ. (2014). Retrieved May 4, 2014, from http://i5.minus.com/ip3ALl7puao0w.png

Rainie, L., Hitlin, P., Jurkowitz, M., Dimock, M., & Neidorf, S. (2012, March 15). The viral Kony 2012 video. *PewResearch Internet Project.* Retrieved September 25, 2012, from http://pewinternet.org/Reports/2012/Kony-2012-Video/Main-report.aspx

razorbeamz. (2014, May 22). LOL IM SO RANDUM CHOOCHOO. Retrieved May 30, 2014, from http://www.reddit.com/r/SummerReddit/comments/267lqd/lol_im_so_randum_choochoo/

readeranon. (2013, November 18). I don't hate female bosses, just my female boss. Retrieved April 11, 2014, from http://www.reddit.com/r/AdviceAnimals/comments/1qwrlt/i_dont_hate_female_bosses_just_my_female_boss/

reddiquette—reddit.com. (2013, April). Retrieved August 1, 2013, from http://www.reddit.com/wiki/reddiquette

reddit.com. (2014, July 14). Retrieved July 14, 2014, from http://www.alexa.com/siteinfo/reddit.com

RedditGifts FAQ. (2014). Retrieved May 10, 2014, from http://redditgifts.com/faq/

Rheingold, H. (2000). *The virtual community: Homesteading on the electronic frontier* (Rev. ed.). Cambridge, MA: The MIT Press.

Riordan, E. (2013, January 30). The Friendzone is a sexist myth. *Feminists-At-Large*. Retrieved April 5, 2014, from http://feministsatlarge.wordpress.com/2013/01/30/the-friendzone-is-a-sexist-myth/

robwinnfield. (2012, November 27). Kid attempts to twerk in his private school uniform to "Birthday Cake" [Comment]. Retrieved July 17, 2014, from http://www.reddit.com/r/cringe/comments/13vi7r/kid_attempts_to_twerk_in_his_private_school/c77rk1m

Romano, A. (2013, April 17). Feminist blogger in hiding after men's rights death threats. *The Daily Dot*. Retrieved May 4, 2014, from http://www.dailydot.com/news/feminist-blogger-in-hiding-mra-death-threats/

Ross, A. (2013). In search of the lost paycheck. In T. Scholz (Ed.), *Digital labor: The internet as playground and factory* (pp. 13–32). New York: Routledge.

Rules of reddit. (2014). Retrieved June 1, 2014, from http://www.reddit.com/rules/

runningman_23. (2011, November 30). Gaming desktop computer. Retrieved May 10, 2014, from http://redditgifts.com/gallery/gift/gaming-desktop-computer/

s.e. smith. (2009a, November 5). Hipster ableism. *Disabled Feminists*. Retrieved June 6, 2014, from http://disabledfeminists.com/2009/11/05/hipster-ableism/

———. (2009b, July 16). Hipster racism. *This Ain't Livin'*. Retrieved June 6, 2014, from http://meloukhia.net/2009/07/hipster_racism/

Salen, K., & Zimmerman, E. (2003). *Rules of play: Game design fundamentals*. Cambridge, MA: The MIT Press.

Salihefendic, A. (2010, November 23). How reddit ranking algorithms work. Retrieved July 5, 2014, from http://amix.dk/blog/post/19588

Samwalter. (2014, April 28). What are people here's opinions on SRS? [Comment]. Retrieved April 30, 2014, from http://www.reddit.com/r/FeMRADebates/comments/246cyx/what_are_people_heres_opinions_on_srs/ch4m16q

Sandstorm_Bot. (2014, May 18). Hi, I'm /u/Sandstorm_Bot. ╭ つ ಠ_ಠ ╲つ DUDUDU Retrieved May 19, 2014, from http://www.reddit.com/r/botwatch/comments/25vni5/hi_im_usandstorm_bot_%E3%81%A4_%E3%81%A4_dududu/

Sandvig, C. (2013). The internet as infrastructure. In W. H. Dutton (Ed.), *The Oxford handbook of internet studies* (pp. 86–108). Oxford: Oxford University Press.

———. (2014, June 26). Corrupt personalization. *Social Media Collective Research Blog*. Retrieved July 16, 2014, from http://socialmediacollective.org/2014/06/26/corrupt-personalization/

Schäfer, M. T. (2011). *Bastard culture! How user participation transforms cultural production.* Amsterdam, the Netherlands: Amsterdam University Press.

Schechner, R. (1985). *Between theater and anthropology.* Philadelphia: University of Pennsylvania Press.

Schensul, J. J., & LeCompte, M. D. (2013). *Essential ethnographic methods: A mixed methods approach* (2nd ed.). Lanham, MD: AltaMira Press.

Scholz, T. (2008). Market ideology and the myths of Web 2.0. *First Monday, 13*(3).

———. (Ed.). (2013). *Digital labor: The internet as playground and factory.* New York: Routledge.

SeaCowVengeance. (2014, January). CompileBot FAQ. Retrieved May 20, 2014, from http://www.reddit.com/r/CompileBot/wiki/faq

secretsafe. (2011, August 21). My dad bought a building and has found a safe behind a hidden wall … opening Tuesday for Reddit. Retrieved July 3, 2014, from http://www.reddit.com/r/reddit.com/comments/jq5io/my_dad_bought_a_building_and_has_found_a_safe/

Senft, T. M. (2008). *Camgirls: Celebrity & community in the age of social networks.* New York: Peter Lang.

———. (2013). Micro-celebrity and the branded self. In J. Hartley, J. Burgess, & A. Bruns (Eds.), *A companion to new media dynamics* (pp. 346–354). New York: Wiley-Blackwell.

Sergnb. (2014, May 2). 30 Rock had a one off character, Jerem, that reminds me of the people I have interacted with on reddit—[0:28] [Comment]. Retrieved May 2, 2014, from http://www.reddit.com/r/videos/comments/24iowz/30_rock_had_a_one_off_character_jerem_that/ch7rmq6

Shefrin, E. (2004). Lord of the Rings, Star Wars, and participatory fandom: Mapping new congruencies between the internet and media entertainment culture. *Critical Studies in Media Communication, 21*(3), 261–281.

Shifman, L. (2013). Memes in a digital world: Reconciling with a conceptual troublemaker. *Journal of Computer-Mediated Communication, 18*(3), 362–377.

———. (2014). *Memes in digital culture.* Cambridge, MA: The MIT Press.

Shirky, C. (2003, February 10). Power laws, weblogs, and inequality. Retrieved March 6, 2013, from http://shirky.com/writings/powerlaw_weblog.html

ShitRedditSays. (2014). Retrieved April 11, 2014, from http://www.reddit.com/r/shitredditsays

Silva, L., Goel, L., & Mousavidin, E. (2009). Exploring the dynamics of blog communities: The case of MetaFilter. *Information Systems Journal, 19*(1), 55–81.

Sirinon. (2014, April 29). What are some interesting secrets about reddit? [Comment]. Retrieved April 29, 2014, from http://np.reddit.com/r/AskReddit/comments/249nej/what_are_some_interesting_secrets_about_reddit/ch50h21

skeen. (2013, June 6). Let's make r/atheism free and open again [Comment]. Retrieved June 30, 2014, from http://www.reddit.com/r/atheism/comments/1fs930/lets_make_ratheism_free_and_open_again/cad9zti

sleepytotoro. (2012, December 18). A Macbook Air (!!!) and so much more! Retrieved May 10, 2014, from http://redditgifts.com/gallery/gift/well-think-won-internet/

smooshie. (2012, February 19). "No information leaves this room": Is Reddit (in danger of) being controlled by an elite few? Retrieved July 20, 2014, from http://www.reddit.com/r/TheoryOfReddit/comments/pwle1/no_information_leaves_this_room_is_reddit_in/

Something Awful. (2014, May 4). *Wikipedia.* Retrieved May 5, 2014, from http://en.wikipedia.org/wiki/Something_Awful

SOPA and PIPA bills lose support on Capitol Hill as Google, Wikipedia and others stage protests. (2012, January 18). *The Washington Post.* Retrieved March 10, 2014, from http://www.washingtonpost.com/business/economy/sopa-and-pipa-bills-lose-support-on-capitol-hill-as-google-wikipedia-and-others-stage-protests/2012/01/18/gIQAwIs38P_story.html

Southern Poverty Law Center. (2012, Spring). Misogyny: The sites. *Intelligence Report.* Retrieved April 5, 2014, from http://www.splcenter.org/get-informed/intelligence-report/browse-all-issues/2012/spring/misogyny-the-sites

SRSTechnology. (2014). Retrieved April 11, 2014, from http://www.reddit.com/r/SRSTechnology

Steinkuehler, C. (2006). The mangle of play. *Games and Culture, 1*(3), 199–213.

SteveTR. (2011). Justin Bieber to North Korea! *Know Your Meme.* Retrieved January 13, 2014, from http://knowyourmeme.com/memes/events/justin-bieber-to-north-korea

Strauss, N. (2005). *The game: Penetrating the secret society of pickup artists.* New York: HarperCollins.

Sullivan, B. (2013, October 24). Competition not found: The rise and inevitable fall of Reddit. *Medium.* Retrieved June 2, 2014

Sullivan, L. L. (1997). Cyberbabes: (Self-) representation of women and the virtual male gaze. *Computers and Composition, 14*(2), 189–204. doi: http://dx.doi.org/10.1016/S8755-4615(97)90020-7

supdunez. (2013, June 4). Trying Out Overly Attached Boyfriend. Retrieved April 11, 2014, from http://www.reddit.com/r/AdviceAnimals/comments/1fonqq/trying_out_overly_attached_boyfriend/

Sutton-Smith, B. (1997). *The ambiguity of play.* Cambridge, MA: Harvard University Press.

Swartz, D. (1997). *Culture and power: The sociology of Pierre Bourdieu.* Chicago: The University of Chicago Press.

Tannock, S. (1995). Nostalgia critique. *Cultural Studies, 9*(3), 453–464.

Taylor, T. L. (2009). *Play between worlds: Exploring online game culture.* Cambridge, MA: The MIT Press.

———. (2012). *Raising the stakes: E-sports and the professionalization of computer gaming.* Cambridge, MA: The MIT Press.

TEDIndia (Producer). (2009, March 10). Alexis Ohanian: How to make a splash in social media [Video]. Retrieved from http://www.ted.com/talks/alexis_ohanian_how_to_make_a_splash_in_social_media

Terranova, T. (2003). Free labor: Producing culture for the digital economy. Retrieved January 1, 2013, from http://www.electronicbookreview.com/thread/technocapitalism/voluntary

TH4N. (2013, June 19). The mods are asleep! Quick, post board games! Retrieved August 1, 2013, from http://www.reddit.com/r/gaming/comments/1gotyz/the_mods_are_asleep_quick_post_board_games/

thanks_for_breakfast. (2012, December 12). 22lb of Idaho potatoes. Retrieved May 5, 2014, from http://redditgifts.com/gallery/gift/22lb-idaho-potatoes/

the_real_stabulous. (2014, July 31). [RECAP] Unibanned! A recap of the fallout of reddit's poster child being banned. Retrieved July 31, 2014, from http://www.reddit.com/r/SubredditDrama/comments/2c9ida/recap_unibanned_a_recap_of_the_fallout_of_reddits/

theempireisalie. (2012, June 17). fumyl figures out how Trapped_in_Reddit "games karma." Retrieved March 10, 2014, from http://www.reddit.com/r/SubredditDrama/comments/v7f1c/fumyl_figures_out_how_trapped_in_reddit_games/

TheLake. (2012, January 26). Meet Omari. Two days ago he returned from the hospital after being hacked in the face by a machete defending an orphanage of 35 children by himself. Think we could raise the $2,000 needed for the remainder of the cement/barbed wire wall to keep both him and the children safe? Retrieved May 14, 2014, from http://www.reddit.com/r/pics/comments/oye34/meet_omari_two_days_ago_he_returned_from_the/

thespamich. (2013, December 4). If 200,000 people die every year from drowning and 200,000 people have already drowned this year, does that mean I can breathe under water? Retrieved May 27, 2014, from http://www.reddit.com/r/shittyaskscience/comments/1s2plu/if_200000_people_die_every_year_from_drowning_and/

Todd, B. (2013, July 16). Does anything go? The rise and fall of a racist corner of reddit. *The Atlantic*. Retrieved April 5, 2014, from http://www.theatlantic.com/technology/archive/2013/07/does-anything-go-the-rise-and-fall-of-a-racist-corner-of-reddit/277585/

Tone argument. (2014, February 19). *Geek feminism wiki*. Retrieved April 11, 2014, from http://geekfeminism.wikia.com/wiki/Tone_argument

Tremayne, M. (Ed.). (2007). *Blogging, citizenship, and the future of media*. New York: Routledge.

Triple Zed. (2013). Scumbag Girl / Scumbag Stacy. *Know Your Meme*. Retrieved April 6, 2014, from http://knowyourmeme.com/memes/scumbag-girl-scumbag-stacy

Tufekci, Z. (2014, March 19). No, Nate, brogrammers may not be macho, but that's not all there is to it. *Medium*. Retrieved May 5, 2014, from https://medium.com/technology-and-society/2f1fe84c5c9b

Turkle, S. (1995). *Life on the screen: Identity in the age of the internet*. New York: Touchstone.

Turner, F. (2006). *From counterculture to cyberculture: Stewart Brand, the Whole Earth Network, and the rise of digital utopianism*. Chicago: The University of Chicago Press.

Turner, V. (1987). *The anthropology of performance*. New York: PAJ Publications.

———. (2001). *From ritual to theatre: The human seriousness of play*. New York: PAJ Publications.

tylercap. (2011, August 16). 2am Chili. Retrieved August 1, 2013, from http://www.reddit.com/r/pics/comments/jkc1j/2am_chili/

UnholyDemigod. (2014). May May June: /r/atheism, and /u/jij vs. /u/skeen. from http://www.reddit.com/r/MuseumOfReddit/comments/21cu8s/may_may_june_ratheism_and_ujij_vs_uskeen/

uniquelyunqualified. (2014, February 26). Gonewild girls confuse me. Retrieved April 5, 2014, from http://www.reddit.com/r/AdviceAnimals/comments/1yzv45/gonewild_girls_confuse_me/

van Dijck, J. (2013). *The culture of connectivity: A critical history of social media*. Oxford: Oxford University Press.

Van Maanen, J. (2011). *Tales of the field: On writing ethnography* (2nd ed.). Chicago: The University of Chicago Press.

VanManner. (2013, February). Internet husband. *Know Your Meme*. Retrieved April 11, 2014, from http://knowyourmeme.com/memes/internet-husband

Vickery, J. R. (2014). The curious case of Confession Bear: The reappropriation of online macro-image memes. *Information, Communication & Society, 17*(3), 301–325.

Vondreau, P., & Snickars, P. (2010). Introduction. In P. Vondreau & P. Snickars (Eds.), *The YouTube reader* (pp. 9–21). Stockholm: National Library of Sweden.

Wajcman, J. (1991). *Feminism confronts technology*. University Park: The Pennsylvania State University Press.

Watson, R. (2011, December 27). Reddit makes me hate atheists. *SkepChick*. Retrieved May 5, 2014, from http://skepchick.org/2011/12/reddit-makes-me-hate-atheists/

———. (2012, October 24). It stands to reason, skeptics can be sexist too. *Slate*. Retrieved May 4, 2014, from http://www.slate.com/articles/double_x/doublex/2012/10/sexism_in_the_skeptic_community_i_spoke_out_then_came_the_rape_threats.html

Weissmann, J. (2014, July 15). Listen as a desperate Comcast rep refuses to cancel a customer's service. *Slate*. Retrieved July 20, 2014, from http://www.slate.com/blogs/moneybox/2014/07/15/comcast_rep_refuses_to_cancel_service_listen_to_the_customer_service_call.html

Wells, P. (2012, November 13). As not seen on TV. *The New York Times*, p. D4. Retrieved from http://www.nytimes.com/2012/11/14/dining/reviews/restaurant-review-guys-american-kitchen-bar-in-times-square.html

Wenger, E. (1998). *Communities of practice: Learning, meaning, and identity*. New York: Cambridge University Press.

whats_hot_DJroomba. (2013, February 20). DAE remember the Digg migration? [Comment]. Retrieved June 2, 2014, from http://www.reddit.com/r/DAE/comments/18xang/dae_remember_the_digg_migration/

White, M. (2006). *The body and the screen: Theories of internet spectatorship*. Cambridge, MA: The MIT Press.

williamshatner. (2013, February 8). Turning off private messages [Comment]. Retrieved April 30, 2014, from http://www.reddit.com/r/ideasfortheadmins/comments/18536m/turning_off_private_messages/c8bocrt?context=

Winkler, A. (2014, March 17). Yes, corporations are people. *Slate*. Retrieved September 24, 2014, from http://www.slate.com/articles/news_and_politics/jurisprudence/2014/03/corporations_are_people_and_that_s_why_hobby_lobby_should_lose_at_the_supreme.html

Winner, L. (1997). Cyberlibertarian myths and the prospects for community. *Computers and Society, 27*(3), 14–19.

WoozleWuzzle. (2012, February 14). "EL OH EL I have you tagged as [_____]" phenomenon because of Reddit Enhancement Suite (RES). Retrieved May 28, 2014, from http://www.reddit.com/r/TheoryOfReddit/comments/ppb6u/el_oh_el_i_have_you_tagged_as_phenomenon_because/

workman161. (2013, April 19). Random act of pizza for the Boston PD, in recognition of their hard in work capturing the Boston Bomber. Retrieved January 15, 2014, from http://www.reddit.com/r/RandomActsOfPizza/comments/1cpoe1/random_act_of_pizza_for_the_boston_pd_in/

yeti-detective. (2013). We need to talk about the Friendzone. Retrieved April 4, 2014, from http://yeti-detective.tumblr.com/post/38349905931/we-need-to-talk-about-the-friend-zone

yishan. (2013, July 17). r/atheism and r/politics removed from default subreddit list [Comment]. Retrieved July 16, 2014, from http://www.reddit.com/r/TheoryOfReddit/comments/1ihwy8/ratheism_and_rpolitics_removed_from_default/cb4pk6g

———. (2014a, September 6). Every Man Is Responsible For His Own Soul. Retrieved September 10, 2014, from http://www.redditblog.com/2014/09/every-man-is-responsible-for-his-own.html

———. (2014b). reddit gold. Retrieved May 18, 2014, from http://www.reddit.com/gold/about

z0mbi3jesuz. (2012). Doom Paul / It's Happening. *Know Your Meme*. Retrieved June 23, 2014, from http://knowyourmeme.com/memes/doom-paul-its-happening

zachinoz. (2014, July 17). I AM Zach Braff. Ask Me Anything. (About Rampart.). Retrieved July 20, 2014, from http://www.reddit.com/r/IAmA/comments/2ayv94/i_am_zach_braff_ask_me_anything_about_rampart/

ZakDougall. (2014, March 2). The girls of GoneWild. Retrieved April 5, 2014, from http://www.reddit.com/r/AdviceAnimals/comments/1zd2v2/the_girls_of_gonewild/

Zuckerman, E. (2013, July 8). Reddit: A pre-Facebook community in a post-Facebook world. *The Atlantic*. Retrieved August 1, 2013, from http://www.theatlantic.com/technology/archive/2013/07/reddit-a-pre-facebook-community-in-a-post-facebook-world/277583/

INDEX

1to34, 135
4chan, 7, 15, 28, 46–47, 56–57, 84, 98, 164
   /b/ board and, 28, 124
   ephemeral experience and, 28
   gaming and, 46
   Poole, C. and, 84
   shitpost and, 118
   summer fags and, 84
4chan Frequently Asked Questions, 118
4chan Rules, 28

A

Aarseth, E., 122
Abad-Santos, A., 48
Abbate, J., 129
About a Boy, 40
About Bit of News, 104
About reddit, 5, 28
activity theory (AT), 25
actor-network theory (ANT), 25

Adam, A., 130
Adamantium, 97
addyct, 86
Advance Publications, 2–3, 29
advice solicitations, 3
Agogino, A.M., 86
AI_Simmons, 43
alanwins, 80
Alexander, J.C., 148
Alfonso, F., III, 135
algorithmic taste profiles, 8
alienth, 145, 163
AlleaGirl, 135
Althoff, T., 34
altruism, 15, 31, 32–35, 37–38, 40, 165
AMAs, 6, 9, 52–53, 64
amlaviol, 55
Andrejevic, M., 7, 32
André, P., 124
Angel, A., 88
anniebeeknits, 33
Ayn Rand what the fuck you say ?!?!?!?!, 107

anti-feminists, 72
anti-SOPA/PIPA efforts, 155
antizeitgeist, 41
antsav888, 80
Attwood, F., 130
AOL and Usenet, 82
AquilaGlobumAncora, 142
Aral, S., 10
Arbitrary Day 2014, 35
ArcheangelleDworkin, 145
ArchangelleSamaelle, 144
ArcheangelleStrudelle, 146
artifact-based platform, 28
askusmar, 36
Associated Press, 49
Attenborough, Sir David, 52
authenticity, 32, 50–56, 64
aux, 140
AyChihuahua, 98

B

BaconReader, 119
Bakhtin, M. and the grotesque body, 20
Banks, D.A., 6
Banks, T., 6, 41
Barbrook, R., 32
Barthes, R., 149
Baum, K., 141
Baym, N.K., 7, 11, 32
BearkeKhan, 134
Beaulieu, A., 11
Believeinfacts, 36
Bennett, W.L., 8
bennyschup, 59
berbertron, 96
Bergstrom, K., 90, 124
Bernstein, M.S., 124
Bianchi, J., 159
Bidgood, 48
Bieber, J., 41, 46
The Big Bang Theory and Sheldon, 58–59
"Big Bang Theory Irony," 59

Bijker, W.E., 2, 25
Bitcoin, 39, 159
Bivens, K., 50
_black, 46
BlackbeltJones, 102
Boczkowski, P.J., 25
Boellstorff, T., 11
Bogers, T., 25
Bonaccorsi, A., 154
Booby_Ooo, 141
Borsook, P., 49
Boston Marathon Bombing, 6, 15, 34, 47, 49, 155
bots (automated scripts), 39, 103–106, 125
  BeHappyBot and, 106
  CaptionBot posts and, 105, 126
  CationBot and, 105–106
  CIRCLEJERK_BOT and, 105
  spam and, 106
  /u/astro-bot and, 104
  /u/autowiki-bot and, 104
  /u/BeHappyBot and, 104
  /u/bitofnewsbot and, 104
  /u/CaptionBot and, 103
  /u/CIRCLEJERK_BOT and, 105
  /u/CompileBot and, 104
  /u/CreepierSmileBot and, 105
  /u/CreepySmileBot and, 105
  /u/DisapprovalBot and, 105
  /u/Extra_Cheer_Bot and, 104
  /u/gandhi_spell_bot and, 104
  /u/GoneWildResearcher and, 133
  /u/haiku_robot and, 104
  /u/Hearing_Aid-Bot and, 104
  /u/Jiffybot and, 103
  /u/MetricConversion-Bot and, 104
  /u/Mr_Vladimir_Putin and, 104
  /u/PleaseRespectTables and, 104
  /u/redditbots and, 104
  /u/Sandstorm_Bot and, 105
  /u/totes_meta_bot and, 104
  /u/VerseBot and, 104
  /u/Wiki_Bot and, 104
Bourdieu, P., 147

Boyd, D., 5
boyd, D.M., 6, 23, 27, 90
Brad, 9, 53, 65, 140–142, 166
Braitch, J., 33
Brandtlyc, 50
BraveSirRobin, 57
BronzeLeague, 119–120
Brown, K., 98
Bruns, A., 50
the bubble, 10
Buchanan, R., 86
Bucher, T., 103
buckybitch, 98
bullying, 42–45
    anti-bullying statement and, 43–44
    disabled individuals and, 43
    inappropriate content and, 42–43
Burrill, D.A., 97

C

Cage, N., 107
Caillois, R., 22, 106
Callon, M., 25
Calloused finger, caring hearts, 33
calmbatman, 82
Captain Picard face-palm, 162
Carmichael, R., 35
Cascading Style Sheet (CSS), 32, 76
    modifications and, 113–114
Catalana, S., 141
Chen, A., 6, 29, 128, 145–146
Chow-White, P.A., 143
Christensen, H.S., 45
Churba, 60
citizen journalism, 8
Clifford, J., 11
Clinton, K., 2, 167
CNN, 49
Colby2012, 81
Coleman, E.G., 24, 28, 45, 97
Coleman, G., 155
collaborative knowledge sharing, 24

colocito, 56
ColtenW, 100
Comes to Reddit Only to Bitch about it, 150
comment gilding and tipping, 39
community, 7, 13, 15–16, 19–20, 22–24, 26, 28, 38, 50–56, 60, 67, 74
    aggregation of, 76
    novice to expert within, 68, 69
    open-source culture and, 154–155
    ownership and, 169
    policing and, 169
    practice of, 68
    sense of ownership and, 85
    technolibertarian leanings and, 154
community platform, 28
computer-supported cooperative work (CSCW), 68
Connell, R.W., 128
Consalvo, M., 22, 122
concerneddad1965, 81
content moderation, 137
content neutral, 164
context collapse, 23
contextual privacy, 52
Coscarelli, J., 145
Costikyan, G., 23
countergaming, 122
    oppositional cultural production and, 122
Creative Commons, 33
creep_creepette, 107
creepshots, 135–136, 150
cringe, 44
    /r/cringe and, 40–41, 43–44
    /r/cringepics and, 40–42
    videos, 41
cringeworthy postings, 15, 40, 42
    not cringeworthy and, 44
crossed, 47
cupcake1713, 89, 153
cyberlibertarian ideals, 49
cynicism, 15, 38, 40–45, 56, 58, 92
    elitism and, 83, 86
    empathy and, 41
    entertainment and, 41

D

Dachis, A., 45
Daddario, A., 133
Danescu-Niculescu-Mizil, C., 34
Daniels, J., 166
Dao_of_Tao, 44
Davis, M., 5
Davison, P., 96
Dawg Diogenes was punk as fuck, 107
definitelymyrealname, 124
Delwiche, A., 7
Democratic People's Republic of Korea (DPRK), 109
denial of autonomy, 138
devicerandom, 110
dhamster, 89
Dibbell, J., 23
Digg, 82–83, 116
Digg.com, 10
digital dualism, 24, 170
Dimock, M., 81
DingoQueen, 133
Dixon Thayer, D., 45
djmushroom, 103
dogecoin, 39, 159
domo-loves-yoshi, 118
Don, 65
Donath, J.S., 102
don't_stop_smee, 57
Down-n-Dirty, 137
doxxing and brigading, 121, 145–146, 155–156, 163
Duggan, M., 49
Dunn, G., 135
During summer, is there actually an increase, 84
Dym, C.L., 86

E

echo chamber, 9–10, 21, 58, 90, 92
EffinD, 96
elite users, 8, 11
Ellison, N.B., 6, 23, 27
EMCoupling, 56
Emerson, R.M., 11
Engeström, Y., 25
Ensmenger, N., 129
Eppink, J., 98
Eris, O., 86
eternal September, 82–84
ethnographic approaches, 11
    loss of distance and, 13
    qualitative researchers and, 11
    reddit and, 13
Explain to me why you just bought reddit gold for Bill Gates, like I am a poor African child about to die from malaria, 109
Explain what is about to happen like I am your dog of 13 years who you are about to have put down, 109
Exquisite Corpse, 98

F

F7u12, 122
Facebook, 6, 10, 24, 27, 29, 36, 41, 43, 52, 69, 81, 102, 125, 139, 162, 167
Facebook Contagion, 167
fapping, 33
fat hate, 44
faq—MensRights, 136
filter bubble, 9
Flanagan, M., 98
flobbley, 97
Foot, K.A., 25
Ford, S., 7
ForWhatReason, 82
Frank_Reynolds_19, 14
Frapstronauts, 33
free information access, 45
free labor, 8
free speech, 45, 162, 164
Fretz, R.I., 11
Frey, D.D., 86

fuckwads, 166
Fung, K., 49
Futurama, 112

G

9Gag, 69
Galloway, A., 122
gatekeepers, 8, 10–11, 53
Gates, B., 35, 52, 109, 142
Gawker, 47, 128
Gawker, ValleyWag, LifeHacker, Jezebel, 47
geek culture, 11, 28, 125, 128–130
    hegemonic masculinities and, 128–130
    identity and, 165
    masculinity and, 61, 97, 137, 150
    sensibility and, 28
gender, 1
genius-bar, 91
G.F., 81
Ghos, R., 5
Gillespie, T., 2, 15, 25
Gillmor, D., 8
girrrrrrrrr1, 80
GitHub, 154, 167
global culture, 2
Goel, L., 5
Goffman, E., 21
goldfish188, 78
goldguy81, 86
Gonewild is a place for open-minded Adult Redditors to exchange their nude bodies for karma, 133
Good Girl Gina (GGG), 140, 143
Google, 6, 24, 27–28, 36, 80, 102, 110, 154, 162
Gottfried, J., 10
Granovetter, M.S., 27
Gray, K.L., 166
Gray, M., 2
Greater Internet Fuckwad Theory (GIFT), 166
Green, J., 7

Greenberg, A., 162
Greenpeace, 46
Groleau, C., 25

H

Halavais, A., 116
hansjens47, 76
Hanson, A., 52
happybadger, 110
Hardin, G., 32
Harrelson, W., 53–54
Harry, D., 124
Haughney, C., 49
Haywood, B., 58
HeLivesMost, 55
Hemsley, J., 81
Henderson, J.J., 7
Henricks, T.S., 22
Hern, A. and the *New Statesman*, 48
Highfield, T., 50
highshelfofsteam, 35
Hindman, M., 154
Hine, C., 11
hipster racism/sexism, 58, 75, 84
hitbart000, 138
Hitlin, P., 81
hivemind, 10, 14, 21, 47, 58, 60, 67, 92, 116, 150, 154
hmasing, 34
Hogan, B., 50, 102
Holcomb, J., 10
homophobic language, 57
hoontur, 123
Horst, H.A., 11
How I Feel When (HFW), 101
Hp9thr, 106
hth6565, 112
Hughes, T.P., 2, 25
Hughey, M.W., 166
Huizinga, J., 22
Huffman, S., 2
Humphreys, S., 166

I

I_am_one_wth_it_all, 55
Iamwoodyharrelson, 54
Ideal Sacrificial Number (ISN) and Neptune, 108
identity, 1–2
Imafidon, 129
image based boards, 24, 28, 80
    /a (anime) and, 28
    /g (technology) and, 28
    /v (video games) and, 28
image macros, 58, 100, 105
    Clarity Clarence and, 134
    College Liberal (CL) and, 142
    Good Guy Greg and, 140
    LOLcats and, 2
    Scumbag Steve and, 139
    Scumbag Reddit and, 58
    Socially Awkward Penguin and, 58, 65
    Unpopular Opinion Puffin (UOP) and, 87
inappropriate behavior, 41
Ingur, 127, 164
Insane Clown Posse and Juggalo, 42
insider/outsider tension, 153
Instagram, 167
Ireallylikepbr, 141

J

JackRig95, 136
jasoneppink, 100
jedberg, 114
Jenkins, H., 2, 7, 47, 167
jkerwin, 82
jmk4422, 149
joserskid, 141
JohnArr, 98
Johnmtl, 36
Jones, S.G., 23
joserskid, 141
jrivers242, 142
judgementbarandgrill, 43

Jurafsky, D., 34
Jurgenson, N., 24
Jurkowitz, M., 81
Juul, J. 22, 120

K

Kang, J.C., 6
Kaptelinin, V., 25
karma, 5, 12–13, 16, 23, 56, 61, 69, 78–79, 87–90, 102, 115, 117–118, 149, 166
    invisible internet points, 115, 120
    mechanics of, 119
    negative and, 69
    redditors and, 118
karmadecay.com, 80
karma jackpot, 118
karma(ic) rewards, 114–117
karma machines, 117
karma points, 32, 56, 80, 116, 118
karmarank, 13
karmawhores and karmawhoring, 16, 23, 78, 79, 115, 117–118
Kelemen, M., 49
Kendall, L., 130, 155–156
    BlueSky forum and, 156
Kerckhove, C.V., 58
Kid Attempts to Twerk in His Private School Uniform to 'Birthday Cake,' 42
KidsAre9532, 59
Kim, B., 105
Kimmel, M., 137–138
Kirkpatrick, D., 52
KittyFooties, 55
kn0thing, 34
Knights of the New, 10
Know Your Meme, 139–140, 142
Knuttila, L., 28
Kollock, P., 23
Konami code, 125
Kony 2012 and http://www.kony2012.com, 81
Kony, J., 81
Kosnik, A.D., 8

L

lacylola, 36
Laina, 140
　See also Overly Attached Girlfriend
LainaOAG, 141
LambdaMOO, 24
Langton, R., 131
Lanier, J., 8
laptopdude90, 104
Lardinois, F., 82
LaTeX_fetish, 140
Latour, B., 2, 25
LavaLampJuice, 139
Lave, J., 68
LeCompte, M.D., 11
leet, 83
Legitimate peripheral participation (LPP), 68
　apprenticeship model and, 68
Leifer, I.J., 86
Lemon, L. and *30 Rock*, 60
Lerman, K., 5
Levy, S., 155
Lewis, H., 129, 157
LBGTerrific, 112
libertarian views, 46
Liebelson, D., 129, 157
Little, Z. and Ridiculously Photogenic Guy, 9
Lion, S., 52
LogicPlacebo, 55
lol/this/first/ttt/etc, 75
Long, D., 34
lukeis2cool, 139
LULZ, 124, 125
Lynn, A., 133, 165

M

MacFarquar, L., 155
MacTheMan, 107
Madrigal, A., 64
magic circle, 22
male gaze, 16, 130–132

Manaka Bros, 44
Marcus, G.E., 11
Marin, A., 141
Markham, A.N., 11
Marlow, C., 5
marriage equality, 46
Marwick, A.E., 9, 23, 43, 90
Mathis, G., 35
Maury Lie Detector, 55, 65
maxgprime, 46
Maxion, 101
McArthur, J.A., 24
McCombs, M.E., 8
McGangbangs, 70
Mcodray, 142
media cultures, 7–11
memes, 3, 24, 26, 53, 79, 82, 84, 96–97, 100, 107, 121, 125
　Business Cat and, 113
　Confession Bear and, 55–56, 75, 105, 113, 121
　OP is a faggot and, 56–58
　/r/AdviceAnimals and, 3, 16, 55–56
　/r/lolcats and, 3
　/r/ffffffffuuuuuuuuuuuu and, 3
　Shiba Inu dog and doge and, 39
　Stormfront Puffin, 56, 65
memetic retelling, 79
men's rights activism, 16
　Equality Canada and, 147
Men's Rights Activists (MRAs), 137, 146–147, 160–161
Messerschmidt, J.W., 128
meta-discourse, 92
Meyer, R., 167
Microsoft, 162
Miettinen, R., 25
mikey_mike24, 92
Miller, D., 11
Milner, R.M., 7, 96
Misa, T.J., 129
misogyny, 45, 75, 127, 133
Mister Splashy Pants, 46
Mitchell, A., 10

MittRomneysCampaign, 147
moab-girl, 42
moderators and moderation, 10, 12, 32, 5, 160
  witch hunts and, 56
Monfort, N., 168
Monroy-Hernández, A., 124
Moore, C., 22
Morozov, E., 8
Morris, K., 124, 146
Mousavidin., E., 5
Mow-bray, 103
Muchnik, L., 10
Mulvey, L., 130
Murray, S., 7
MuscleT, 142
Museum of the Moving Image, 98, 100
Mylaptopisburningme, 34
My Reaction When (MRW), 101

N

Naaman, M., 5
Nahon, K., 81
Nakamura, L., 143, 166
Nardi, B.A., 11, 25, 123
Neidorf, S., 81
NetLingo List of Chat Acronyms & Text Shorthand, 83
net neutrality, 45–46
neologisms, 87, 92
neuroticfish, 43
new media, ephemera of, 2
New York Post, 49
Nice Guy Syndrome, 133–134
niche discussions, 7
  MetaFilter and, 7
  Slashdot, 7
niche interests, 3, 85
  /r/bicycling and, 3
  /r/hiphopheads and, 3
  /r/MakeupAddiction and, 3
  /r/PenmanshipPorn and, 3

Nicholson, J., 133
Nicole, K., 46
Nine Things to Know about RedditGifts, 35
Nissenbaum, H., 52
non-participation (np) links, 91
Norris, P., 9
Norton, M.I., 166
not real redditors, 165
not-safe-for-life (NSFL), 52, 132
not-safe-for-work (NSFW), 16, 41, 51, 74, 76, 132–136
  attention whores, 134
  Gone Wild (GW) and, 133–134
  homosexuality and, 42
  /r/gonewild (GW) and, 51, 133–135, 138
  /r/NSFW_GIF and, 133
  /r/pornvids and, 133
NovaXP, 39
novelty accounts, 102–103, 106, 111, 125
  /u/AWildSketchAppears and, 102
  /u/cupcake1713 and, 160
  /u/DiscussionQuestions and, 102
  /u/GradualBillCosby, and 102
  /u/IcanLego, and 102
  /u/JudgeWhoAllowsThings and, 102
  /u/KimMyungKi and, 109
  /u/NedTheCosmicManatee and, 102
  /u/PigLatinsYourComment and, 103
  /u/r_spiders_link and, 111
  /u/Shitty_Watercolour and, 102
  /u/SingsYourComment and, 102
novelty subreddits, 106–111
  collective ethos of reddit and, 107
  /r/birdswitharms and, 107
  /r/catReddit and, 107
  /r/explainlikeIAmA and, 108–109
  /r/explainlikeimfive (ELI5) and, 3, 108
  /r/fourthworldproblems and, 110
  /r/fifthworldproblems and, 109–110
  /r/fuckingphilosophy and, 107
  /r/harlemshake and, 107
  /r/infiniteworldproblems and, 110
  /r/MURICA and, 107
  /r/onetrueGod and, 107

/r/photoshopbattles and, 107
/r/Pyongyang and, 109
   American hegemony and, 109
/r/sixthworldproblems and, 110
/r/shittyask[blank] and, 107
/r/shittyask-science and, 107
/r/subredditoftheday and, 112
/r/switcheroo and, 112
/r/TheStopGirl and, 107
/r/thirdworldproblems and, 110
Nussbaum, M., 131, 138
Nutshapio, 82

O

Obama, Barack (American president), 52, 69
Ohanian, A., 2, 3, 31, 86
   Without Their Permission: How the 21ˢᵗ Century Will Be Made, Not Managed and, 31
OKCupid, 41
O'Neill, S., 35
online infrastructure, 45
On the Theme of Higher Education Haters, 55
Ooer, 86
oops777 and *The Atlantic Wire*, 48
open-source idealism, 154–156
open-source technology, 45, 164
opspe, 122
O'Reilly, T., 5, 26
original content (OC), 77, 79–80, 101
original poster (OP), 10, 33, 43–44. 51, 53, 55, 80, 91, 101, 117–118, 123
   bundle of sticks and, 58
   expectations and, 57
   policing behavior and, 57, 74
   self-posts and, 79
outsider status, 61
OverjoyedMuffin, 84
Overly Attached Girlfriend (OAG), 140–141

P

Panovich, K., 124
Pariser, E., 9
participatory culture, 1–2, 7, 15–16, 28, 167–169
   affiliation and, 2
   critique of, 1
   mediated and, 2, 14
   optimism and, 31
participatory culture platform, 5, 7, 15, 27–28, 68, 74, 103
   commmodified and designed spaces and, 31
   Digg delicious and, 5
   Metafilter and, 5
   open-source nature and, 85
   politics and, 2, 14–16, 165–167
Pascoe, C.J., 57
patterned interaction, 91
Paul, R., 46
Pearce, C., 11, 22, 123
Pedogeddon, 146
Pekhota, 86
Penny Arcade, 166
pertnear, 134
Pew's Internet & American Life Project, 81
Phillips, W., 90, 124–125
Pinch., T., 2, 25
Pinterest, 167
piracy, 45
pk_atheist, 137
PlantSomeTrees, 53
plasmatron7, 146
play, 1, 15–16, 20, 22–24, 50, 111, 119, 148, 165
   collective activity and, 123
   gaming spaces and, 123
   inverting and, 122
   patterns of, 95
plexico_mcbean, 107
Popper, B., 49
Postigo, H., 122
power user model, 10

privacy, 45
private messages (PMs), 5
progressive attitudes/causes, 46
project8an, 139
pro-privacy stance, 43
prosumer of content, 7
Protect IP Act (PIPA), 45–46
prototypical redditor, 64
　See also geek culture
pseudoanonymity, 50–56
public figure humiliation, 41
Punamäki-Gitai, R.-L., 25
pun and pile-on threads, 97–98
Purushotma, R., 21, 167

R

rageguy face, 122
/r/all, 3, 12, 33, 39, 42, 55, 64, 70–71, 76–77, 84–85, 87, 97, 133, 140, 150, 160, 161, 163
Rainie, L., 81
Raja, T., 129, 157
Rand, M., 141
/r/antiSRS FAQ, 146
razor-beamz, 83
reaction GIFs, 24, 57, 98–102, 119, 121
　Sweet Brown and, 100
　upvote/downvote and, 101
readeranon, 142
the_real_stabulous, 117
recap flair, 113
reddit, 1–7, 17, 19–20, 28, 164
　ableism and, 58
　acculturation and, 68
　activism and mobilization and, 6, 45–47
　　social issues and, 47
　administrator accountability and, 169
　algorithmic sorting and, 63, 167
　aggregator of links/platform and, 3, 19, 26
　altruism/cynicism dichotomy and, 38, 49–50, 67
　anything goes space and, 6

carnival and, 15, 20–21, 124, 154, 165
child pornography and, 121, 127
circlejerk and, 67, 70, 81, 92
conspiracy theorists and, 160
creativity and, 165
culture and, 22, 32
detectives and, 47–50
discourse and, 16, 149
downvotes and, 75–76, 114–117, 119, 148, 154
front page of the internet and, 2
homophobia and, 58, 71
humor and, 53–54, 58, 70, 81, 90, 107, 165
　racist, sexist, or classist statements and, 165
identity and, 23, 24, 51, 57, 58, 60, 61–63
　diversity and, 61, 63
　integrity and, 52
insider perspective and, 85
level of disclosure and, 54
link-sharing/news-sharing site and, 19, 25–26, 60, 76, 85
lurking and, 11–12, 69
mob justice and, 47–50
mob mentality and, 55, 68
norms and, 54
patterns of interaction, 85
performance/ritual and, 15, 20–21, 92, 148
phenomenological experience and, 19
platform and, 20, 24–25, 28, 60, 63, 111, 125, 165, 167–168
　logics and, 85
　See also participatory culture platform
politics and, 63, 76, 114
popular culture and, 58, 84, 90, 97
pseudoanonymity and, 24, 28, 90, 119, 168
racism and, 58, 71, 75–76
reputation and, 119–120
rules and, 16, 23, 57
search engine and, 80
self-loathing and, 59

sexism and, 71, 76, 127
    objectification of women and, 130
sexual harassment and, 134
sharing of information and, 121
skepticism and, 80
snoo and, 58
social justice perspective and, 14
social news/social sharing and, 25–26
spam and, 121
specialness and, 54, 86
the real-name/one-identity policies and, 168
upvotes and, 75–76, 114–117, 119, 148, 154
voting behavior and, 119–121
reddit bananaroo and, 112
reddit celebrities and, 115, 117, 119
reddit.com, 5, 161
reddit culture, 13–14, 32, 84, 165
    collective sense of self and, 85
    gift economy as, 32–33
Reddit Enhancement Suite (RES), 101, 111–113, 119–120, 126
    tagging and, 111–113
reddit-friendly retailers, 39
RedditGifts, 28, 35–39, 46, 51, 163
    Elf and, 37
    exchange credits and, 36
    gallery and, 36
    project and, 37
    redditgifts.com and, 35
RedditGifts FAQ, 36, 63
Reddit Gold, 5, 28, 39, 63, 106, 109, 115, 142, 162
    trophy case and, 27, 39
reddit history, 77
reddit platform, 4, 77
    logics and, 12
redditor, 4, 15, 68, 70, 75, 156
    collaborators, 54
    lowest common denominator, 70
    newbie and, 68
    statistics on, 10
Redditor's Wife (RW), 141–142

reddiquette, 12, 17, 73–77, 122, 154
    guidelines and, 74
    individualistic rationality and, 77
    posts and, 76
    rationality over emotionality, 75
reddiquette–reddit.com, 74, 121
reddit rules, 120–122
reddit switcheroo, 111
    Old Reddit Switcharoo and, 112
repost policing, 79–82
reposts, 83
Rheingold, H., 23–24, 168
    the WELL and, 24
Richards, A., 129, 147
Riordan, E., 133
Robison, A.J., 2, 167
robwinnfield, 42
Romano, A., 147, 157
Rose, K., 141
Ross, A., 8
Rossi, C., 154
Rules of reddit, 121

S

Sagan, C., 79
Salen, K., 22, 121
Salihefendic, A., 10, 114
Samwalter, 146
Sandstorm_Bot, 105
Sandvig, C. 25, 29
Sarkeesian, A., 129, 147, 153
Schaaf, A., 127
schadenfreude, 41
Schäfer, M.T., 122
Schechner, R., 2, 148
Schensul, J.J., 11
Scholz, T., 8
Schrute, D. and *The Office*, 60
Schwarzenegger, A., 35, 53
science-and-technology studies (STS), 25
Scumbag Stacy, 138–142
SeaCowVengeance, 104

Senft, T.M., 9, 130, 147
Sergnb, 60
Shatner, W., 39, 152
Shaw, D.L., 8
Shaw, L.L., 11
Shefrin, E., 7
Shifman, L., 1, 96
Shirky, C., 13, 23
shitposts and shitposting, 16, 83, 117–118
ShitRedditSays (SRS), 12, 16, 144, 160
    BRD mascot and, 147
    counterperformance and transgressive play, 147–150
    polarizing entity and, 144
    rational discourse and, 144
    safe space and, 152
    SRS shill and, 161
    subversion and, 147
    succession and, 147
shitty community network, 108
    /r/shittyama and, 108
    /r/shittyfoodporn and, 108
    /r/ShittyPoetry and, 108
    /r/shittyprogramming and, 108
    /r/shittysocialscience and, 108
Silva, L, 5
Sirinon, 145, 150
skeen, 79
Skepchick, 127, 146
slacktivism, 45
Smith, A., 49
Smith, M., 23
Smith, s.e., 58
Smith, W., 49
smooshie, 11
SMS messaging, 41
Snapchat, 167
Snickars, P., 29
social justice warriors (SJWs), 160–161, 165
social networking model, 27
social networking platforms, 8
social networking sites (SNSs), 23, 27
Something Awful forums, 146, 157
Sommers, S.R., 166

SOPA and PIPA Bills Lose Support, 45
Soundcloud, 6, 27
Southern Poverty Law Center, 137
space of emergence, 22
spectatorship, 90
SRSTechnology, 144
Steinkuehler, C., 22
STEM, 64, 86
SteveTR, 46
The Stop Online Piracy Act (SOPA), 45–46, 127
Stratton, A., 139
    See also Scumbag Stacey
Strauss, N., 137
strength of weak ties, 27
subreddits, 4, 10, 12–14, 19
    anti-SRS and, 146
    /r/AngryBrds and, 146
    /r/antisrs and, 146
    /r/SRSSucks and, 146
    connection with others and, 33, 38
    defaults and, 4, 16–17, 32, 70–73, 96, 122, 153–154, 160
    Cakeday posts and, 117
    ethic ands, 45
    fatphobic and, 44
    flair and, 113–114
    guidelines and, 122
    meta- and, 16, 21, 46, 58, 63, 70, 83, 84–87, 90, 92, 159
    circlejerk and, 114
    meta-conversations and, 165
    niche and, 62
    online Secret Santa and, 35
    /r/AcademicPhilosophy and, 71, 77
    /r/AdviceAnimals and, 3, 55, 83, 87, 96, 113, 132, 134, 139–141
    /r/analogygifs and, 100
    /r/apple and, 123
    /r/AskAcademia and, 113
    /r/AskHistorians and, 9, 96, 114
    /r/AskReddit and, 3, 51, 59, 81, 83, 92, 106, 132, 145, 150, 161
    /r/askscience and, 3, 9, 107

INDEX

/r/Assistance and, 34
/r/atheism and, 79, 81, 146
/r/atheismrebooted and, 79
/r/aww and, 3, 76, 98, 117
/r/babyelephantgifs, 3
/r/bicycling and, 3
/r/books and, 35
/r/Braveryjerk and, 114
/r/Calligraphy and, 19
/r/CandidFashionPolice (CFP) and, 135, 164
/r/catreactiongifs and, 100
/r/changemyview and, 114
/r/chicago and, 50, 75, 113, 122
/r/circlebroke and, 14, 85, 101, 119, 144
/r/circlejerk, 46, 64, 81, 87–90, 92, 114, 149–150
    Dadaist space of play and, 150
/r/classicrage and, 122
/r/creepshots and, 6, 127, 135–136
/r/creepy and, 72
/r/diabetes and, 33
/r/dogecoin and, 39
/r/dogetipbot and, 39
/r/Drama and, 85
/r/depression and, 104
/r/EatingDisorders and, 33
/r/fatlogic and, 44
/r/fatpeoplehate and, 44–45
/r/FeM-RAdebates and, 145
/r/firstworldproblems and, 109
/r/fffffffuuuuuuuuuuuuu and, 3, 122
/r/findbostonbombers and, 47–50
    See also Boston Marathon Bombing
/r/Fitness and, 53
/r/food and, 70, 72
/r/FreeKarma and, 114
/r/funny and, 72–73, 75, 84, 97, 117
/r/gameoftrolls and, 124
/r/Games and, 121, 122
/r/gaming and, 87, 122–123
/r/Gamingcirclejerk and, 87
/r/gentlemanboners and, 132
/r/gifs and, 76, 98, 100–101

/r/girlsinyogapants and, 136, 155
/r/GreatApes and, 6, 153, 164–165
/r/HighQualityGifs and, 100
/r/hiphopheads and, 3
/r/IAmA and, 3, 6, 9, 52, 108
/r/incest and, 164
/r/IncestPorn and, 153
/r/jailbait and, 146
/r/JusticePorn and, 34
/r/KarmaConspiracy and, 115
/r/LadyBashing and, 146
/r/lolcats and, 3
/r/loseit and, 33
/r/lounge and, 39
/r/MakeupAddiction and, 3, 35
/r/melodicdeathmetal and, 19
/r/MensRights (MR) and, 72, 136–138
/r/mildlyinteresting and, 72, 78
/r/moderatorjerk and, 87
/r/movies and, 35
/r/Museumof Reddit and, 85
/r/Music and, 78
/r/niggers and, 138
/r/NoFap and, 33, 36
/r/northkorea and, 126
/r/OutoftheLoop and, 85
/r/outside and, 125
/r/pcmasterrace and, 90
/r/PenmanshipPorn and, 3
/r/philosophy and, 71
/r/pics and, 72, 97–98, 117, 133
/r/PicsOfDead-Kids and, 138, 153
/r/Pizza and, 91
/r/Polandball and, 86
/r/politics and, 76
/r/popping and, 132
/r/programming and, 82
/r/progresspics and, 33
/r/random and, 110
/r/Random_Acts_Of_Amazon (RAOA) and, 34
/r/Random_Acts_Of_Pizza (RAOP) and, 34
/r/reactiongifs and, 100–101

/r/relationships and, 33, 51, 57
/r/SampleSize and, 12
/r/science and, 96
/r/SecondWorldProblems and, 110
/r/seduction and, 136–137
/r/selfharm and, 33
/r/sex and, 51, 132
/r/shitpost and, 118
/r/ShitRedditSays (SRS) and, 85, 87, 143–147
   circlejerk and, 143
   FemPire and, 144, 153, 160–161
   /r/SRSFunny and, 144
   /r/SRSGaming and, 144
   /r/SRSMythos and, 145
   /r/SRSTechnology and, 144
/r/showerbeer and, 19
/r/spacedicks and, 153
/r/StarWars and, 117
/r/SubredditDrama (SRD) and, 43, 58, 85, 87, 91–92, 113, 160–161
   No drama rule and, 161
   popcorn and, 90
   popcorn pissing and, 91
   standard dramafests and, 91
/r/SubredditDramaDrama and, 91
/r/subredditoftheday and, 148
/r/SuicideWatch and, 33
/r/SummerReddit (SR) and, 82–83
/r/TheFappening and, 162, 164
/r/TheoryofReddit (TOR) and, 10, 84–86
/r/TheRedPill (TRP) and, 136–138, 160
/r/trees and, 52
/r/TrollXChromosomes and, 100–101
/r/TumblrInAction and, 160
/r/TwoXChromosomes (2X) and, 71–72, 77, 86, 160
   period shits and, 160
/r/Unidan and, 9, 52, 117, 110
/r/upvotegifs and, 101
/r/videos and, 71, 74, 76, 138
/r/WatchItForThePlot and, 133
/r/WTF and, 132
/r/WhatsInThisThing and, 57
/r/WhiteRights and, 153
/r/worldnews and, 104
/r/writingprompts and, 72
/r/WTF and, 77–78, 161
/r/YogaPants and, 136
social support and, 33
themed exchanges and, 35
subreddit drift, 77–79
   poor moderation and, 79
   See also moderators and moderation
Sullivan, B., 82
Sullivan, L.L., 130
summer friends, 83
Summer Reddit, 83
supdunez, 141
Sutton-Smith, B., 22
Swartz, A., 155
Swartz, D., 147–148
System Management Unit (SMU), 123

T

Tannock, S., 20
Taylor, S.J., 10, 97
Taylor, T.L., 11, 22, 24, 123
technology and relationships to, 25
TEDIndia, 46
teens/preteens, 41
Terranova, T., 8
text-based responses, 57
TH4N, 123
thanks_for_breakfast, 36
theempireisalie, 80
TheLake, 34
thespamich, 108
throwaway account, 51
TinyURL, 74
Tits or GTFO, 134
toaster studies, 2
Todd, B., 138
Tone Argument, 144
Tremayne, M., 50

Tripathi, S., 48
Triple Zed, 139
trolls and trolling, 5, 16, 71–72, 75,
     124–125, 160
   the mask and, 124
Tufekci, Z., 150
Tumblr, 74, 98, 160, 167
Turkle, S., 23, 168
Turner, F., 24, 49, 155
Turner, V. 21
Twitter, 10, 15, 27, 43, 74, 125, 162, 164, 167
tylercap, 96
Tyson, N. d., 52, 71, 79

U

Un, Kim Jong, 109
UnholyDemigod, 79
uniquelyunqualified, 134
Unobtainium, 97
/u/Trapped_In_Reddit, 80

V

values, 45
van Dijck, J., 8
Van Maanen, J., 11
VanManner, 141
Vargas, G., 124
Vered, K.O., 166
Vickery, J.R., 55
Violentacrez, 47, 145–146, 155
Vimeo, 6
Vondreau, P., 29
vote brigading/manipulation being
     another, 125

W

Wajcman, J., 128–129
Watson, R., 127, 146–147, 153, 157

Web 2.0, 5–6, 8, 23, 26–27, 167
   authenticity and, 50
   brand and, 43
   gatekeeping and, 7–11
   neoliberalism and, 8
web platforms and commodified and
     designed spaces, 31
Weigel, M., 2, 167
Weismann, J., 88
Wells, P., 70
Wenger, E., 68
Wernersen, R.N., 25
whack-a-mole, 163
Whats_hot_Djroomba, 83
Wheaton, W., 35
White, M., 90–91
white knights, 165
WikiLeaks project, 155
williamshatner, 152
Winkler, A., 162
Winner, L., 49
Wong, Y., 28, 162
Wordpress, 167
work for the text, 7
workman161, 34
World of Warcraft game, 118
   /r/wow and, 118
WozzleWuzzle, 111

Y

yishan, 39, 162
yeti-detective, 133
Youtube, 6, 10, 15, 27, 29, 43, 74, 104, 140,
     167

Z

z0mbi3jesuz, 46
Zimmerman, E., 22, 121
Zuckerman, E., 6, 23
Zuckerberg, M., 52

General Editor: **Steve Jones**

Digital Formations is the best source for critical, well-written books about digital technologies and modern life. Books in the series break new ground by emphasizing multiple methodological and theoretical approaches to deeply probe the formation and reformation of lived experience as it is refracted through digital interaction. Each volume in **Digital Formations** pushes forward our understanding of the intersections, and corresponding implications, between digital technologies and everyday life. The series examines broad issues in realms such as digital culture, electronic commerce, law, politics and governance, gender, the Internet, race, art, health and medicine, and education. The series emphasizes critical studies in the context of emergent and existing digital technologies.

Other recent titles include:

Felicia Wu Song
  Virtual Communities: Bowling Alone, Online Together

Edited by Sharon Kleinman
  The Culture of Efficiency: Technology in Everyday Life

Edward Lee Lamoureux, Steven L. Baron, & Claire Stewart
  Intellectual Property Law and Interactive Media: Free for a Fee

Edited by Adrienne Russell & Nabil Echchaibi
  International Blogging: Identity, Politics and Networked Publics

Edited by Don Heider
  Living Virtually: Researching New Worlds

Edited by Judith Burnett, Peter Senker & Kathy Walker
  The Myths of Technology: Innovation and Inequality

Edited by Knut Lundby
  Digital Storytelling, Mediatized Stories: Self-representations in New Media

Theresa M. Senft
  Camgirls: Celebrity and Community in the Age of Social Networks

Edited by Chris Paterson & David Domingo
  Making Online News: The Ethnography of New Media Production

To order other books in this series please contact our Customer Service Department:
  (800) 770-LANG (within the US)
  (212) 647-7706 (outside the US)
  (212) 647-7707 FAX

To find out more about the series or browse a full list of titles, please visit our website:
  WWW.PETERLANG.COM